Oil, Turmoil, and Islam in the Middle East

Oil, Turmoil, and Islam in the Middle East

Sheikh R. Ali

PRAEGER

New York
Westport, Connecticut
London

Library of Congress Cataloging-in-Publication Data

Ali, Sheikh Rustum.
    Oil, turmoil, and Islam in the Middle East.

    Bibliography: p.
    Includes index.
    1. Petroleum industry and trade—Political aspects—
Near East. 2. Near East—Politics and government—
1945–   . I. Title.
HD9576.N36A58   1986   338.2'7282'0956   85-31254
ISBN 0-275-92135-2   (alk. paper)

Library of Congress Catalog Card Number: 85-31254
ISBN: 0-275-92135-2

First published in 1986

Praeger Publishers, 521 Fifth Avenue, New York, NY 10175
A division of Greenwood Press, Inc.

Printed in the United States of America

The paper used in this book complies with the Permanent
Paper Standard issued by the National Information Standards
Organization (Z39.48-1984).

10 9 8 7 6 5 4 3 2 1

To Rina, Bappi, and Bobby

Contents

1

Introduction

Oil produced in the Middle East is critical to the needs of the industrialized world. Oil is the key energy source fueling economic growth. This dependency on Middle East oil will intensify unless the United States and other industrialized nations commit a sizable expenditure over the next several decades to revamp our energy-consuming infrastructure or find alternative energy sources.

The current world demand for oil amounts to about 20 billion barrels annually. The United States alone consumes over one-fourth of this amount. The Middle East's capacity to supply a significant part of the world's oil is due to the area's unique geology, endowing it with a large concentration of giant and super-giant oil fields. The Middle East has about 60 percent of the world's oil reserves. The economic power inherent in this concentration of a vital resource creates a great deal of political power.

The Arab countries have about 333 billion barrels of oil in reserve, or about 52 percent of the world's total. Much has been written and said, correctly and incorrectly, about the influence of Arab oil on the Arab-Israeli conflict in particular and on world politics in general. The supposition that Arab oil power can and will influence the United States to accept Arab territorial demands and impose them on Israel is based largely on the assumption that the Arabs are in a position to wage economic warfare successfully on the United States and its allies. This assumption is derived from the efforts of the 1973–74 Arab oil embargo. Diplomatic intervention by the United States and the shape of the two Sinai agreements prove the efficacy of the Arab oil weapon potential.

The oil embargo continues to affect strategic thinking greatly. It is therefore vitally important that the embargo be carefully studied if policymakers are to judge the variables which may or may not resolve Arab-Israeli differences.

A review of intra-Arab dialogue before, during, and after the embargo shows that political realities in the Arab world—particularly the rivalry between progressive and conservative governments—doomed the oil weapon from the start. In an analysis of the application of the embargo, it is evident that Western multinational oil companies were able to insulate the impact of the embargo from its principal target. A critical review of world oil production and consumption, as well as economic relations between the primary actors during 1973–74, demonstrates monopolistic and monopsonistic forces which all but canceled the theoretical bases of the oil weapon's effectiveness.

The Middle East is of utmost importance to the world for at least two reasons. One is oil, the other Islam. The two may be considered separately or as a unit. Oil is important as the source of wealth and weaponry from which the political clout of the region is derived. This, in turn, creates much of the present turmoil in the Middle East. Islam, the ever-present and most dominant religion of the region, is a powerful force that unites the region against outsiders and directs the lives of most Middle Eastern peoples. Together, oil and Islam symbolize life in the Middle East, past and present.

It is unsettling that since the 1970s some Middle Eastern regimes thought to be among the strongest and most stable—Iran under the shah, for example—have turned out to be quite fragile. Others, such as Jordan, that have been characterized as weak or lacking in popular support and therefore on the verge of collapse, have been remarkably durable. The shattering of the surface calm of some of the oil-rich nations—for example, Saudi Arabia and Kuwait—by events such as the storming of the Grand Mosque in Mecca in 1979 and the explosion in the U.S. Embassy compound in Kuwait in 1983 provides substance to suspicions that wealth and economic prosperity do not translate directly into stability.

Islam has a guiding and often decisive influence on the conduct of governments and individuals. It is significant that the non-Islamic world has belatedly discovered the Islamic revival only after Iranian militants sought the late shah's return to face Islamic justice. That the shah's supplanter seems to have succeeded by appealing to religious sentiment has astonished Western observers, who take it as axiomatic that religion and politics have little or nothing to do with one another.

The principal means by which Muslim nations of the Middle East have asserted themselves is an oil oligopoly. The oil cartel has been remarkably effective in directing attention to the displacement of the Palestinians, with all its implications for relations between the affluent industrial world and the primary producers of the Southern Hemisphere.

Against this backdrop an analysis of the importance of Middle Eastern oil in world politics will be made. Any consideration of the forces influencing the pattern of development in the Middle East should take Islamic tradition into account.

ARAB AND ISLAMIC OIL WEAPON

Middle Eastern oil is, in a sense, Arab oil, because, with the exception of Iran, all oil producers in the region are Arabs. It is also Islamic oil since all Middle Eastern oil producers, including Iran, are Muslim nations. The concept of Arab nations' using oil as a diplomatic weapon against the West and Israel is as old as the Arab-Israeli conflict itself.

Of the four Arab-Israeli wars so far, oil figured significantly in all but that of 1948. In 1956, when Britain and France attacked Egypt, Israel joined in against Egypt, making it the second Arab-Israeli war. The Arab countries not only blocked the flow of oil to Israel, but also to the Mediterranean. The Suez Canal and the pipelines from Iraq and Saudi Arabia were closed. The situation arising from the closure of the Suez Canal and the embargo on Arab oil shipments to Britain and France produced serious economic consequences. The embargo increased oil prices in Britain and France. The oil companies were forced to obtain the commodity from the United States, which shipped their Middle Eastern oil products via the Cape of Good Hope. That, too, entailed more cost. As a result, oil rationing was introduced in Western Europe.

In 1967, when the third round of the Arab-Israeli conflict began, Middle East oil became even more important to most of the industrial world because oil had almost replaced the use of coal since 1965. When war broke out, then Egyptian President Gamal Abdel Nasser announced that Britain and the United States had joined Israel in its attack on the Arabs.[1] He also said that the Sixth Fleet had helped Israel attack Egyptian airports and military bases. These accusations were not true, but at the time the Arabs took them seriously. Notices were issued to oil companies to cease exporting oil to Britain, the United States, and other countries blacklisted by the Arabs. Germany was added to the list because it sold gas to Israel. For the first time Arab oil-producing nations shut down all operations. In further retaliation in Saudi Arabia and Kuwait, "bands of saboteurs assembled with explosives, ready to destroy the big companies' installations once and for all."[2] Before the oilfields could be blown up by the saboteurs Arab troops moved in. Soldiers were ordered to stop the flow of oil, and their presence prevented sabotage. The pattern of the 1956 oil embargo was repeated. The Suez Canal was closed once again, this time indefinitely.

The embargo lasted only two weeks and was relatively inefficient. There were several reasons for the embargo's failure. First, because of inaccurate propaganda, the Arabs decided to use their oil weapon as the only alternative to inaction. Second, within a few days they discovered that they lacked the financial strength to carry on the oil embargo without getting money from the oil companies. "Saudi Arabia was the first to feel the pinch acutely. . . . King Faisal was informed by his Finance Minister that there was no more money in the till, and that for once Aramco was unable to help."[3]

The world's oil industry changed in the beginning of the 1970s with the disappearance of surplus oil in the United States and the growing control of multinational oil companies by the Organization of Petroleum Exporting Countries (OPEC). The United States and other industrialized noncommunist countries of the world were dependent on the supply of oil from the Middle East. Oil industries in the United States stressed the significance of Middle East oil, especially that of Saudi Arabia, as important to the Western Hemisphere.

When the war broke out on October 6, 1973, it looked at first as though the Arabs would not need to use the oil weapon. When the war turned against the Arabs the oil ministers of the Organization of Arab Petroleum Exporting Countries (OAPEC) met in Kuwait on October 17 to consider the role of oil in the Arab struggle to liberate the lands occupied by Israel in 1967. In their attempt to bail out Egypt and Syria, which were directly engaged in the war against Israel, the Arab oil ministers decided that each country would immediately cut oil production. October 17 was a red-letter day in the history of the world energy crisis. All Arab countries joined in the war against Israel by imposing an immediate 5 percent cut in oil production and by raising that cut monthly by 5 percent until Israel withdrew from the occupied Arab territories.

Unlike the production cutback agreement, the OAPEC oil embargo against the United States, the Netherlands, and other nations was imposed individually and outside the OAPEC organizational structure. As a result, the oil embargo lacked the common institutional mechanism crucial to the effective supervision and implementation of collective, common policies.

An analysis of the government statements of the countries participating in the embargo reveals that the OAPEC members did not have the same perception of the objective they were pursuing nor did they have a common understanding of the principles that govern the political uses of oil. There are a number of ways to illustrate this point.

One example would be Iraq's refusal to honor the Kuwait production cutback agreement. Saddam Hussein, then vice-president of Iraq's Revolutionary Command Council, explained the Iraqi decision in the following way:

> The policy of reducing production generally harmed other countries more than America ... the production cutback in the form encouraged by reactionary circles means effectively depriving the countries of Western Europe and Japan of their oil requirements and causing them an extremely serious crisis.[4]

Hussein's explanation demonstrates a different understanding of the tactical use of the oil weapon rather than of the overall strategy. The OAPEC decision presupposed that the most effective use of the weapon was against the West in general, while Iraq's decision was predicated on the theory that it was best utilized selectively. For some leaders, like Libya's Muammar Qadhafi, it was in the Arabs' interest to "hit America by striking Europe." In other words, the Kuwait decision was based on the theory that by crippling Western Europe, the North Atlantic Treaty Organization (NATO) members would be forced to take one of two initiatives: first, to pressure the United States to change its Middle East policy or, second, which Qadhafi preferred, to ally themselves "with the Arab world and Africa against Washington and Moscow."[5] In Qadhafi's view, the latter would have been "the most desirable, the most logical and wisest for Europe."[6]

The theory of pressuring the United States through its European allies was considered dangerous by Iraq. Iraq feared that by causing a European economic crisis, the Arabs could jeopardize if not destroy Europe's growing independence from the United States. As Hussein explained:

> In recent years thanks to the independent political path followed by General de Gaulle, Western Europe has gradually begun to turn toward a position relatively independent from America on many issues, among them the Arab cause.... This development is in the interests of the Arabs who should attempt by every means to develop this attitude and build Arab-European relations and reciprocal advantage.[7]

For Iraq, therefore, a nondifferentiated oil cutback policy would adversely affect trends that Iraq perceived as ultimately benefiting relations between the Arab world and Western Europe. Iraq saw the European weakness as a political danger to the Arab cause. In short,

> the occurrence of a suffocating economic crisis in Western Europe ... may drive them to issue relatively favorable statements now, but those countries will find themselves forced, during the next phase, to abandon their independent policies and the greater part of the production of the Arab countries.[8]

OIL PRICE

From 1973 to 1974, OPEC's price-setting mechanism essentially quadrupled world oil prices.[9] A freeze on oil prices was agreed on by OPEC in December 1974. A 10 percent price increase was announced in September 1975. After maintaining Saudi Arabia's imposed price freeze for most of 1976, OPEC members split in December, with Saudi Arabia and the UAE raising oil prices 5 percent. The remaining members raised prices by 10 percent, with a further 5 percent rise to come in mid-1977. In December 1978, a further price increase of

14.5 percent was announced. The next price hike on June 28, 1979, was between 16 and 24 percent.[10]

The most fractious meeting in OPEC's history came in December 1979. The cartel failed to agree on any uniform price. However, a temporary compromise was reached in Vienna in September 1980. Under this agreement the price of Saudi oil rose $2 per barrel to $30, while other OPEC members froze their prices at existing levels, which averaged about $32 per barrel. In October 1981 OPEC fixed a new united base price of $34 per barrel and froze it through the end of 1982. During 1983 a highly significant event took place in world oil markets. Following price cuts by Britain and Norway, both non-OPEC members, Nigeria, a major OPEC member, broke ranks with OPEC and unilaterally reduced its oil price. For the first time OPEC officially announced a decrease in its benchmark price for oil. The official price of $34 per barrel was reduced to $29. In addition, OPEC adopted a production ceiling of 17.5 million barrels.

In 1984, for the second consecutive year, the behavior of the international oil markets favored consumers. At an emergency meeting OPEC decided to hold the line of posted prices by reducing total production quotas to 16 million barrels per day (b/d), and the price was fixed at $28. In July 1985, an OPEC meeting in Vienna to prop up prices collapsed.

EFFECTS OF PETRODIPLOMACY

What has been an energy crisis for most of the world has been for the oil producers a tremendous political and economic bonanza. The relatively successful Arab military action in the October war of 1973 and Arab use of oil as a weapon showed that they were advancing not only in the tactics of modern warfare, but also in the art of diplomacy. The world's moderation has been facing pressure from those in the Middle East who feel from experience that militancy pays.[11] It was for this reason that even pro-Western Saudi Arabia shunned its policy of moderation and decided to use the oil weapon in the Middle-East conflict. This satisfied Arab nationalist aspirations. At the same time, oil producers realized their weapon was a two-edged sword. Before the weapon could cause irreparable damage to the Western economy which, in turn, would affect Middle Eastern economic interests, the oil embargo was eased.

The early lifting of the embargo indicates that the oil producers avoided abuse of their newfound economic power. "There is a growing realization that this power based on hydrocarbons is finite and that alternative sources of revenue should be developed speedily."[12]

As time passed, the effect of Arab petrodiplomacy was felt more in the international economic field than in the political arena. This happened as a result of oil price escalation for which the Arabs cannot be blamed. Although Saudi Arabia led the Arab nations in imposing the embargo, it is doubtful that they intended to increase prices so suddenly. Nevertheless, the situation got out of control and, taking advantage of the oil embargo, Iran and Venezuela took the lead in escalating the price.

The growing disparity between the oil haves and have-nots has brought about international financial chaos and misunderstanding. This inequality has been accentuated by the fact that international oil is owned by a few developing nations which are still poor compared to Western standards. Naturally, they jealously guard their newfound wealth against outside encroachment.

Not all jealousies and conflicts lend themselves to a quick solution. Even though OAPEC and OPEC are not yet convinced, no return to the old order is possible. A transfer of financial power, with all its ramifications, has already taken place. The world must adjust to this new reality, as the oil producers call it.

As long as OAPEC and OPEC remain united, they can do whatever they like. Historically speaking, all international cartels have contained the seeds of their own destruction. The oil cartels cannot be the exception. Already serious conflicts of interest among member nations are evident. Because producers are many, some competitive forces may operate to break the unity. However, the oil-producing countries are now acting more independently. If this trend continues the oil cartels will exist as "paper tigers" only.

TURMOIL

The Iranian revolution, riots in Egypt, civil war in Lebanon, and the seizure of the Grand Mosque in Mecca all occurred during the 1970s. In some form or other this turmoil constitutes a challenge combining religious discontent with social, tribal, ethnic, and ecopolitical opposition to the establishment in particular countries.

Since the Iranian revolution, the Middle East has suffered more unrest than any other part of the world. In April 1983, a suicide-truck bomber attacked the U.S. Embassy in Lebanon causing the deaths of 63 people. Six months later, on October 23, two massive explosions in American and French command posts in Lebanon killed 241 U.S. Marines and 59 French paratroopers. Less than two weeks after this combined assault, another suicide bomber struck the Israeli military post in the Lebanese port of Tyre, killing 61 people.

Events soon proved that this new terroristic tactic was not to be confined to Lebanon. On December 12 a dump truck packed with explosives was driven into the U.S. Embassy compound in Kuwait, killing 5 and wounding 37.

In September 1984, yet another suicide attack took place at the U.S. Embassy annex in Lebanon, killing 12 and wounding 60. As recently as June 1985 a Shiite movement called Amal (Hope) hijacked a TWA flight in Beirut, killing one passenger and releasing others after 17 days of detention.

Terrorism is a form of political violence, and it is a fact that the United States is the favorite terrorist target in the Middle East. Although the French, Saudis, Kuwaitis, and others have also been targeted, 200 of approximately 695 terrorist attacks from 1983 to December 1985 were against the United States.[13]

In these attacks, organizations like the Islamic Jihad (Holy War), Amal, or Hizbullah (Party of God) usually claim responsibility. Iranians are accused by the Western news media and American officials of belonging to these groups of terrorists.

The Iranian Shiite revolution is intent on propelling its Jihad throughout the largely Sunni Middle East. Whether it is able to overthrow the governments of the Persian Gulf states largely depends on the outcome of the six-year-old Iran-Iraq war. In the Middle East, governments adhering to the more moderate Sunni Islam are becoming more strict in enforcing religious observance in an attempt to insulate themselves from the radical fundamentalist Shiite influence.

ISLAM

Islam, the largest and most influential religion in the developing Afro-Asian world, remains relevant to politics, but in different ways.

"Indeed Islam is politically crucial today to a degree that Christianity has not been since the age of the Crusades or the Reformation."[14]

Depending on conditions, Islam can fuel a revolution or serve as an anchor for individuals or societies in times of flux. It is a vibrant and dynamic religion—a way of life. It has never been dormant. Islamic resurgence has occurred typically in response to crises. That is exactly what is happening in the Middle East where the Iranians, for example, are seeking salvation from Western and Communist ecopolitical and cultural domination through religious revivalism.

The pattern of Iran, whose Muslims are overwhelmingly Shiite (a minority sect in the world), is not duplicated in other countries in the region. In Saudi Arabia the ruling authorities are allied with the religious establishment. In Iraq the events in Iran have tended to sharpen the antagonism between the Sunni (the majority sect in the world) elements and the Shiite community. Despite these differences the recovery of Islamic identity in Iran is being felt throughout the Muslim world as a response to the dilemmas and disillusionments of the time and the frustrations in socioeconomic life.

The four Arab-Israeli wars since Israel's creation in 1948 have inspired much soul-searching within the Arab and Islamic worlds. The thrust of the rally, with Islam as a focus of identity and universal faith encompassing many developing nations, has been to bring together some 850 million followers. The call to rally to the Palestinian cause, for instance, adds impetus to the Islamization of the Arab-Israeli conflict.

One major characteristic of Islamic recovery is its potential as a unifying movement. However, because of the segmented structure of most Middle Eastern countries along ethnic, linguistic, and sectarian lines, the tendency toward Islamic unity has often produced significant intercommunal conflict. Despite the new Islamic fervor, Islamic solidarity is not a generator of potent collective action in world or regional politics. Today, the most politically powerful ideology in the world is nationalism. The revival of Islam in countries such as Iran is more significant in supplanting the force of nationalism.

It seems profitless to speculate on what an Islamic revival in the Middle East would mean to the rest of the world. It is possible, of course, that Islam might assume the role of arbiter of state policy in many Muslim countries.

The trouble and turmoil in the Middle East raise serious questions about the supply of oil to the outside world, particularly to the industrialized world of the West and Japan. Occasionally the oil producing nations threaten to cut off oil supplies to industrialized nations, usually on political grounds. The visible pro-Western states are now showing reluctance to maintain an uninterrupted supply of oil and are imposing new conditions. The political situation in Saudi Arabia, domestic turmoil in Iran, regional conflict in the case of the Iran-Iraq war, and the perpetual Arab-Israeli hostilities may curtail or shut off oil exports. Even with inter-Arab relations and intra-Arab rivalries deemed important by oil-producing nations, there may be partial or total embargoes on exports. Among other possibilities affecting the supply of oil are acts of sabotage by terrorists, civil war, and regional war. Any of these factors may boost the price of oil.

NOTES

1. Leonard Mosley, *Power Play: Oil in the Middle East* (Baltimore: Penguin Books, 1974), p. 343.

2. Ibid.

3. Ibid., p. 344.

4. Hussein's statement in *Al-Thawra* (Baghdad), December 20, 1973.

5. *Le Monde* (Paris), October 29, 1973.

6. Hussein's statement in *Al-Thawra*, op. cit.

7. Ibid.

8. Ibid.

9. Jack C. Plano and Roy Olton, *The International Relations Dictionary* (Santa Barbara, Calif.: ABC-Clio, 1982), p. 142.

10. *New York Times*, June 29 and July 4, 1979.

11. *Middle East Economic Digest* (London), March 22, 1974.

12. Ibid., December 28, 1973.

13. United States Department of State, Bureau of Public Affairs, *Terrorism in the Middle East* (Washington, D.C., October 1, 1984), and ABC television, *World News Tonight with Peter Jennings*, December 31, 1985.

14. Michael Walzer, "The Islam Explosion," *New Republic*, December 8, 1979, 181, no. 23, pp. 18–21.

2

International Energy
and Oil Politics

Energy has been a dominant influence in international politics and has wrought significant changes in the overall relationship between industrial and raw-material-producing countries. Energy is the economic lifeblood of nations and is central to the whole complex of international economic relations. On the political front, the problem of supply, demand, and price has compelled energy-importing countries to find new ways of dealing with major oil-exporting nations. The narrowing gap between global energy resources and their consumption not only threatens our modern way of life, but our very survival.

During the 1970s the world finally recognized the international energy crisis. The Western countries and Japan realized that energy and security were linked. The Arab oil embargo of 1973-74 brought to a climax a trend which had begun developing much earlier: a gradual shift from European and American coal to cheaper but unreliable Middle-East oil. The embargo drove energy consumers to find newer and better ways to offset future oil supply disruptions and to reduce dependence on imported oil through conservation and increased use of alternative fuels such as coal, nuclear power, and solar energy.

The abundant availability of energy resources, along with technical ability and economic competition may, in the long run, provide sufficient energy to meet world needs. In the immediate future the drive toward energy self-sufficiency in the United States—the world's largest energy consumer—through conservation, recovering oil from onshore and offshore areas, and various other means has reduced America's dangerous dependency on imported oil.[1]

Dramatic changes in world energy markets have been a striking, but largely unpublicized, evolution of the way industrialized nations define the notion of international energy security. The turning point in the evolution of energy security came at the 1981 Ottawa summit.[2] That meeting of seven industrialized nations marked the beginning of a new awareness among the major energy importers that nonoil sources of energy imports no longer be necessarily considered better or more secure than the oil imports they wanted to replace. At Ottawa, the United States argued that serious energy security risks could also arise from excessive dependence on other imported fuels, particularly natural gas. What concerned the American delegation was the construction of the new Soviet gas pipeline to Western Europe. President Ronald Reagan himself urged European nations to consider the security implications of this project and to examine indigenous alternatives to an increased dependence on natural gas.

Since the Ottawa summit, the Western dialogue on energy security and natural gas has been strained by differing perspectives on the Siberian pipeline and U.S. sanctions on the sale of oil and gas equipment. The Soviet pipeline controversy dramatized the basic differences between American and Soviet policymakers in the guise of a dispute over energy supplies. By insisting that the issue be viewed in an East-West context of confrontation and cold war, U.S. officials politicized it in a way that sharpened differences between the United States and Western Europe. Claiming that Western Europe would become dependent on Soviet gas, the Reagan administration chose to forget that members of the North Atlantic Treaty Organization (NATO) were already buying oil and gas from the Soviet Union, which they consider to be both competitive with and more dependable than Middle Eastern oil. To the European NATO partners the pipeline imbroglio showed, in the words of a U.S. scholar:

First, the futility of unilateral sanctions that required multinational implementation; and second, the gulf between American conceptions of energy security based on broader economic determinants.[3]

In retrospect, it does not seem that the Soviet pipeline created problems for Western security, but the oil embargo and the accompanying production cut put the security and economic well-being of the West and Japan in jeopardy. In response to this security problem, an instrument of cooperation among the industrialized nations in energy matters was developed.

INTERNATIONAL ENERGY AGENCY

In 1974, at the height of the energy crisis, Secretary of State Henry Kissinger first proposed an international energy program to coordinate the energy policies of industrial oil-importing nations. Thirteen foreign ministers of the major oil-consuming nations gathered in Washington on February 11 for a three-day conference on energy policy. The nations represented at the Washington energy conference were Belgium, Canada, Denmark, France, the Federal Republic of Germany, Ireland, Italy, Japan, Luxembourg, the Netherlands, Norway, the United Kingdom, and the United States. Among the proposals the participants discussed was a seven-point program of cooperative action presented by Kissinger. In effect, the communiqué issued at the end of the conference incorporated much of the Kissinger proposal: (1) cooperation on the conservation of energy; (2) development of alternative sources of energy; (3) research and development; (4) sharing of oil in emergencies; (5) international financial cooperation; (6) an invitation to developing countries to join the major consumers of oil; and (7) cooperation between consumers and producers.[4]

The United States, not being as dependent on Arab oil as many others, could follow an independent, pro-Israeli policy in the Middle East. On the other side of the Atlantic, U.S. allies were being humiliated by having to assume, under pressure, a pro-Arab policy in order to maintain their oil supplies. The disarray in the NATO alliance was evident when, at the height of the fourth Arab-Israeli war, Kissinger launched his shuttle diplomacy from Washington to Moscow and the Mideast capitals without consulting U.S. allies in Western Europe. The Atlantic allies refused to permit U.S. aid to Israel from bases within their territories. This disorder in NATO's rank and file encouraged the Arabs to believe that an oil embargo against the nations supporting Israel would produce public and diplomatic opinion against Israel, as it did in Europe. The Kissinger strategy in convening the conference, aside from creating energy-sharing proposals, had its political significance in renewing confidence in the NATO military alliance.

As expected, oil producers vigorously objected to the conference. The Arab oil producers, particularly, were disturbed by the far-reaching cooperative efforts being made to create an international energy agency to deal with such emergencies as embargoes by sharing available oil, cutting consumption, and using stocks on an equitable basis. Kissinger's global energy program elicited muted response from most Arab capitals.[5] Nevertheless, the conference marked the beginning of the renewal of confidence in the Atlantic alliance under U.S. leadership.

The conference participants agreed to establish a coordinating group to direct its actions. France protested this and other declarations of the conference. France saw in the conference initiated by the United States a device by which to reassert its economic domination of Europe.

The United States and other industrial countries met in Brussels to give concrete shape to the proposed international energy agency. The Kissinger plan was still undergoing modifications. Ultimately, what emerged was a four-point program of international cooperation that Kissinger hoped would help avert disruptions of the Western and Japanese economies. The United States, Canada, Japan, Turkey, and 12 European countries formally created the International Energy Agency (IEA) in Paris on November 15, 1974. The four main features of the IEA, or counter oil cartel, as critics call it, are: (1) the establishment of an energy-sharing agreement among the major consumers in the event of a new embargo; (2) the formation of cooperative conservation and energy-development programs; (3) the establishment of a $25 billion fund to recycle petromoney into deficit countries; and (4) the convening of a conference among producer and consumer countries.[6]

The energy-sharing, stockpiling, and conservation programs are now a part of IEA. The creation of a recycling fund has been organized through the International Monetary Fund (IMF).

THE CONTINUING CRISIS

After the first energy crisis of 1973–74 the world began to cope with the situation. Unfortunately, in 1979 Iranian supplies were interrupted and the world was taught another lesson. This was a clear example of the West's vulnerability to attack from an unexpected source. The IEA was designed to meet a recurrence of the 1973 experience—a major shortfall (above 7 percent) or an embargo. But history did not repeat itself. The interruption in 1979 was considerably less than 7 percent.

The industrialized world soon learned that even a small interruption, clouded by uncertainty, could have devastating economic consequences. Although the decline in Iranian production caused only a 4 percent decrease in world production, market dislocations and price increases buffeted a world economy which had still not fully recovered from the 1973 oil price hike. These blows pushed the world into a recession and slowed economic growth in the industrial nations. The 1979 price hikes ultimately caused a loss of about $300 billion of the industrialized nations' gross national product.[7]

The IEA developed a response to the 1979 crisis. Its members made a commitment to reduce the group's demand for imported oil by 2 million barrels per day (b/d). Each nation's share of the reduction, as well as the time within which action had to be taken, was unspecified. Although savings of about 1.5 million b/d, or 6 percent of 1978 IEA imports, were achieved by the end of 1979,[8] it was too late to forestall sharp price increases. The IEA members, focusing on the modest shortfall, had underestimated the psychological impact of the crisis on the market.

At the June 1979 Tokyo summit, the seven largest industrial nations adopted 1985 oil import targets. The IEA substantially refined and expanded these into 1980 national import ceilings for all its members. This process was institutionalized by the establishment of a system within the IEA in which ceilings were imposed to counteract a market shortfall. Each nation was responsible for determining the measures necessary to achieve its own ceiling.

The industrialized nations were faced with another threat—the Iran-Iraq war. The energy effects of the war between the two OPEC members were serious, although manageable. The war took 3.9 million b/d of oil exports off the market, over 8 percent of that produced in noncommunist countries. Since consumption declined, the world could simply do without some of this oil—about 1 million b/d of it. An additional 1–1.5 million b/d was made up through increased production from OPEC and non-OPEC members.

Although the primary threat to the energy security of the United States, Western Europe, and Japan remains their continued vulnerability to oil supply disruptions, there have been a number of changes in the world energy situation since the oil supply disruptions of the 1970s that have significantly strengthened world energy supply.

For instance, the United States and its allies have diversified their sources of foreign oil. During 1983, U.S. oil imports from Mexico and Canada were more than double its imports from the Persian Gulf. Other IEA countries also have diversified their sources of foreign oil and, since 1979, have decreased their dependence on OPEC oil from 40 percent to about 28 percent of their energy requirements.[9]

Parallel to the decline in energy consumption in the United States since 1973, the rate of energy consumption per dollar of GNP is estimated to have declined by 15 percent in Europe and by slightly less than 10 percent in Japan. During the same period, oil consumption is estimated to have declined by more than 10 percent in Europe and by almost 15 percent in Japan. Despite these efforts, dependence on imported oil remains high in many countries. Therefore, further reducing vulnerability and strengthening emergency response measures continue to be primary international concerns.

The international energy picture during 1984 and 1985 was characterized by unusual stability. As a result of significant worldwide declines in oil consumption over the years, oil supplies are plentiful and prices have remained relatively constant or declined. These declines can be attributed to a combination of relatively weak economic growth in the industrialized countries, increased substitution of alternatives to oil, such as coal, solar, and nuclear energy, and the continuing adjustment to the 1973–74 oil markets completed in 1981.

Although conditions in the world oil market currently are favorable, those conditions are subject to change. The United States has taken a number of measures to protect against market disruptions, and is encouraging its allies to take similar actions. Since 1981, the amount of oil stored (buried in a salt cavern) in the Strategic Petroleum Reserve (SPR) in West Hackberry, Louisiana, increased from 110 million barrels to 503 million barrels. This represents about 100 days of imports at the current level. Efforts are being made to increase the size of the SPR to 750 million barrels. In addition, and in response to market forces, U.S. oil companies have decreased imports from insecure sources in the Middle East and North Africa and have increased supplies from secure sources such as Britain, Norway, Canada, and Mexico. About 30 percent of U.S. oil imports in 1983 came from OPEC, compared to 70 percent in 1977.[10]

THE IRANIAN THREAT

The Iran-Iraq war is now in its sixth year. During this period, the political chaos in Iran has given rise to a cacophony of conflicting voices in its foreign policy, which aggravated Iran's relations with the outside world. Equally disturbing for the Middle East is Iran's commitment to export its revolution to the region. Iran has been broadcasting its revolutionary ideals to the conservative gulf states. Furthermore, Iran has repeatedly threatened to close the Strait of Hormuz to Iraqi and other foreign vessels; it is through the strait that about 8 million barrels of oil pass daily to the noncommunist world.

In that event the United States might be pressured to prevent Iran from closing the strait. President Reagan reiterated at a news conference on February 22, 1984, that there was "no way that we could allow that channel to be closed."

The threat of U.S. military intervention in the 26-mile-wide Strait of Hormuz is believed to be real because the United States has been occasionally undertaking desert maneuvers since the 1973 Arab oil embargo. Although there is disagreement among State Department and Pentagon policy planners, the latter are reportedly keeping their contingency plan up-to-date and military maneuvers seem to be continuing.

The industrialized nations appear to be ready to deal with a sudden loss of Persian Gulf oil. They are better equipped to face a disruption now than they were in 1973–74, when they were caught by surprise. Also, consumption has dropped and oil production outside the Middle East has risen. If the strait is closed, the United States, which now imports less than 5 percent of its oil from the Middle East, could rely on the early sale of SPR oil to minimize the effect of the disruption. Hardest hit may be Western European nations, which depend on the Persian Gulf supply for 28 percent of their oil, and Japan, which gets half of its oil from the gulf.

ALTERNATIVE SOURCES OF ENERGY

While the industrialized world has come a long way in redefining energy security since the 1981 Ottawa summit, it is more difficult to predict the future of energy security. Just as many Western policymakers no longer assume that nonoil imports are more desirable than dependence on foreign oil, it is also clear that they no longer believe that the energy policy inexorably and automatically moves as though immune to market forces. The positive consequence of this realization has been that energy policymakers have learned to concentrate on a basic economic relationship and fundamental requirement, that is, on the marketplace. Market forces, combined with the structural changes in world energy markets brought about by frequent and precipitous oil price hikes, caused the slack global energy market and the lower energy prices we see today.

The down side of these changes is that the need for continued energy security is seen by many as less compelling. On one hand, we are aware that energy markets are likely to tighten considerably when the world economy rebounds. This strengthening is already occurring to some extent. Such a trend augurs renewed attention to energy security issues. On the other hand, working in conjunction with the psychology of complacency, the same economic downturn that was responsible for the declining world energy demand and falling prices has also restricted the financial capacity of the industrialized nations to expand energy-security efforts. Budgetary stringencies and weak demand have brought drawbacks to energy security measures, calling into question the timely development of indigenous energy alternatives.

Nuclear Fuel

Enough study has been done to permit a reasonable assessment of the energy sources that today appear ample. The first in the series to be considered is the production of electricity from nuclear fuels. In any discussion of nuclear fuel the public reacts as though to the danger of the atomic bomb. As many as 60 independent scientists have concluded in a study done for the reorganized U.S. Atomic Energy Commission (AEC) that worries about nuclear power have been vastly exaggerated.[11] James T. Ramey, a former commissioner of the AEC, maintains that today's reactors feature enough built-in safeguards to prevent accidents.[12] Nuclear power is two-faced; it is clean and does not pollute the environment as fossil fuel does, but its radioactive waste is dangerous for many lifetimes. The social cost of nuclear power is being measured. The U.S. government has invested over $1 billion in an attempt to measure the environmental and social costs of nuclear power.[13] As a result, a vast amount of literature has appeared, fueling public debate on nuclear hazards. The debate is unbalanced in the absence of any parallel assessment of the danger of burning fossil fuels. The issue can hardly fail to be resolved as more studies on the danger of burning fossil fuels appear.

Many scientists believe that the fuel of the future is hydrogen. An innovation in this field is the use of hydrogen to fuel the internal-combustion engine. American astronauts have used hydrogen power cells for electrical power in their spacecraft. Scientists at the Jet Propulsion Laboratory in Pasadena, California, have successfully demonstrated an automobile engine that runs on a mixture of hydrogen and gasoline. Its operation is clean, meeting most federal emission standards set for automobiles.[14]

Another group of U.S. scientists has developed an engine that converts gasoline into hydrogen. Pilot studies are being conducted to provide pure hydrogen for propulsion. These innovations, along with the attempt to use hydrogen for heating and cooling homes, may not only bail the IEA countries out of the energy shortage, but save the world from future economic peril. Hydrogen-powered automobiles may be mass-produced within ten years.

Concern over nuclear power has increased due to the incident at Three Mile Island in Pennsylvania in 1979. Unless there is a major accident, nuclear power may rapidly grow in the United States, as it has in the Soviet Union, France, and other European nations. A uranium shortage could force a switch to breeder reactors if production of thermal reactors peaks after 1985. The large-scale use of nuclear power in the late 1980s is likely to release oil and natural gas for more pressing uses, particularly for automobiles. There are over 200 nuclear power plants now operating in the United States.

The near disaster at Three Mile Island represents the most serious nuclear accident in the United States. Nuclear opponents contend that a full-fledged disaster was narrowly averted and that it demonstrated the inherent dangers of nuclear power. Proponents argue that the danger has been unduly sensationalized.

The consequences of the accident were profound. Governments all over the world nervously reassessed their nuclear power plants. A presidential commission in the United States concluded that although no one had been harmed, the accident was extremely serious, and it recommended a restructuring of the Nuclear Regulatory Commission (NRC). Sweden scheduled a referendum on nuclear power. Japan closed its nuclear plants. In the United States and the United Kingdom, vocal opposition to nuclear power was expressed through lobbying by antinuclear and environmental groups. Public demonstrations against nuclear power often broke into violence in West Germany.

Despite opposition, both the West and the Soviet Union steadily develop nuclear energy. In the United States, President Ronald Reagan allowed the construction of a demonstration 350-megawatt fast-breeder reactor at Clinch River, Tennessee, work on which had been stopped by President Jimmy Carter. The United Kingdom, West Germany, Austria, Sweden, France, and the Soviet Union announced new plans for developing more fuel from nuclear power.

Coal

Our selections for possible development in the near future include synthetic crude oil (syncrude) and synthetic gas (syngas). Like nuclear energy, these selections are not made arbitrarily but are based on technical feasibility and resource availability with financial consequences that the industrial economy can bear.

The United States alone has 3.2 trillion tons of coal deposits. Therefore, with proven oil deposits being quickly depleted, coal seems to promise a fuel-based transportation system. Syncrude, syngas, and methyl alcohol (methanol) can be produced from coal. Although there is great uncertainty in predicting the cost of new technology, the Massachusetts Institute of Technology Energy Laboratory Policy Study Group notes that for all processes involved in this new technology the costs are very similar to that of the old technology.[15] The National Petroleum Council (NPC) had projected that by 1985 syngas would account for 2.5 trillion cubic feet of gas per year "at the maximum rate physically possible without any restrictions due to environmental problems, economics, etc."[16]

The NPC predicted that the demand for gas in 1985 would be 41 trillion cubic feet; the supply of domestic gas would fall short by 10 trillion cubic feet. To fill this gap with syngas would have required the construction of at least 120 coal gasification plants before 1985 at a cost of $24 billion.

Research and development for syncrude has been slower than for syngas. According to one study, the United States could have had ten syncrude plants by 1985, each with a 100,000-b/d capacity.[17] The oil shortage in the United States sparked new interest in coal liquefaction technology, since coal is the largest energy resource of the United States. By producing syncrude from coal, the United States could maintain a 150-year supply.[18] This is in addition to coal's other uses.

The foundation for development of syngas and syncrude may be laid in this decade. Although no studies or data are available now, it is possible that during the decade from 1985 to 1995 syngas and syncrude will replace natural gas and oil to some extent. Syngas was once widely used in the United States.[19] It was about 1955 when the utilities in the United States shifted from syngas to natural gas. This can be reversed without hardship on the economy or on consumers. Similarly, new syncrude technology can be developed without major hardship within a decade.

Solar Energy

Many scientists and engineers have begun working on methods of converting solar energy into heat, electricity, and chemical fuels. Solar water heaters have been in commercial use in California and Florida since the 1930s.[20] They are also in use in many European countries, Japan, the Soviet Union, Australia, and Israel. Solar radiation is so abundant that the energy arriving from the sun on just 0.5 percent of the United States is more than enough to satisfy U.S. energy requirements until the year 2000.[21] Because of the cost and difficulties of installing solar equipment in existing homes and because of the slow rate of replacing housing, the technology of solar heating and cooling will take place several decades before it could have any significant impact on energy use. There are various projections on this. Scientists at the American Association for the Advancement of Science feel that because of the growing shortage of fuels, solar energy will be used widely within several decades. Scientists have suggested that as society begins to realize that cheap fossil fuels are not easily available, more solar panels will appear on rooftops throughout the United States. Solar panels are already in use throughout the world. Their widespread use in existing homes is a real possibility as conventional energy sources dwindle and the political climate hitherto inimical to development of solar energy disappears.

FUTURE LESSONS IN ENERGY SECURITY

International energy security cannot be separated from international economic policy, international security policy, or from international relationships as a whole. They are inexorably linked. The oil crises over the past decade made that amply clear.

Looking back at the energy crises of the 1970s, OPEC has emerged as a superpower in setting oil production quotas, distribution, and prices. While this group of formerly indigent nations has gained almost absolute power in deciding oil policies, the major consuming nations have lost their ability to influence the course of events in energy matters as well as in world affairs. The industrialized Western nations and Japan have lost their control over the market and exposed themselves to the real and potentially disastrous consequences of supply disruptions.

The problems are complex and critical. But, acting together, the industrialized nations can probably minimize their impact. In this regard, there are at least five lessons to be digested. They are: (1) reliance on market forces and a more market-oriented approach to international energy security; (2) stockpiling, or maintaining a strategic reserve of oil sufficient to withstand a short-term supply interruption in the future; (3) cooperation among energy-importing nations; (4) long-term efforts aimed at reducing dependence on energy imports; and (5) determination to pursue an international environment conducive to stable world trade.

In summary, it is the responsibility and duty of industrialized nations to take a reasoned, long-term look at their energy security, mindful of, but not overwhelmed by, the economic and financial difficulties now prevailing.

INTERNATIONAL OIL POLITICS

The big concern in the 1970s was that skyrocketing oil prices would bring about a worldwide recession, and rising prices did contribute greatly to plunging the world into the feared recession. Now there is a fear of the repercussions resulting from declining oil prices. The drop in prices would be an extremely serious problem if it forced oil-producing countries to default on their debts. The situation does not look promising. OPEC oil production is expected to be less this year than last year. Because of the special nature of the oil industry, there is the danger that this could bring about a collapse of the pricing system.

The decade of the 1970s witnessed an extraordinary rise in the power of oil-exporting countries, which had been eroding the control of international oil companies and slowly boosting prices toward the dramatic increases of 1973.

The most significant shift in control, however, was the expansion of direct marketing of oil by many members of OPEC. Until this shift, the transnational companies or the so-called seven sisters dominated oil business in the world.[22] Since the Arab oil embargo of 1973, many state-owned enterprises within producer nations have increased the share of oil marketed directly to consumers. Directly marketed oil rose from 8 percent of trade in 1973 to 42 percent in 1985.

The continuing worldwide recession, coupled with energy conservation and the consequent oil glut, served to keep prices down throughout 1983. This created severe strains within OPEC. Last year, there were several price cuts by both OPEC and non-OPEC nations. Many emergency meetings broke up in disagreement as members fought over production and price ceilings. It is not surprising that the continued existence of OPEC as an effective body is being called into question.

Persistent slack oil demand and soft prices have led to both pleasant and unpleasant developments. Non-OPEC nations began to produce and sell more oil at competitive prices to the hitherto almost monopolistic OPEC market. Oil sources shifted dramatically from OPEC to non-OPEC, which provided 50 percent of the world's oil. Increased oil revenue in the hands of OPEC has helped to finance highly ambitious, in some cases unrealistic, development plans. It has also increased luxury spending, which has led to new problems. Nigeria, a leading OPEC nation, almost went bankrupt last year but was ultimately saved by Saudi Arabia which offered loans. The decline of oil prices also brought Mexico, a major non-OPEC oil producer, to the brink of economic collapse, from which it was rescued by the United States. In return for this friendly gesture, Mexico agreed to deliver $1 billion in additional oil for the U.S. Strategic Oil Reserve.

After ten years of unchallenged supremacy over the oil market and with freedom to continually raise prices, the 13 members of

OPEC began battling among themselves over what to do about the precipitous drop in prices and the weakening demand for oil. OPEC, staggered by the threat of a global price war, agreed on March 14, 1983, to cut its price nearly 15 percent and to limit each member's oil production. Besides dropping the benchmark price from $34 to $29 per barrel, OPEC set an overall production limit of 17.5 million b/d. This was the first time in its 23-year history that the OPEC cartel reduced prices. OPEC agreed to its unprecedented reduction after Britain, an important non-OPEC oil exporter, decided earlier to slash its oil prices by $3 per barrel. Still, many OPEC ministers believe that the price cut is temporary and that reduced production and lower prices will quickly eliminate the oversupply.

To save face and to show an outward unity, OPEC members reached the price and production-cut agreement after a prolonged meeting in London. But not all OPEC members can abide by the agreement for long. Members in desperate need of funds are likely to break ranks by undercutting quotas and dropping prices below those agreed on. Indeed, within five days of the London pact, Iran ignored the $29 benchmark price and sold its oil at $26 per barrel. Other nations like Algeria, Nigeria, Indonesia, Iraq, and Venezuela may also sell oil below the official price to maintain their development plans and to support their large populations. The biggest threat to OPEC's price stability may come from outside. The Soviet Union earns an estimated 60 percent of its currency from its oil exports. It sells 40 percent of its oil on the spot market. The Soviets take whatever price they can get.

The oil crisis has given way to an overabundance. The glut, popularized in the news as indicating the success of conservation and use of alternative energy sources, is probably a transitory phenomenon. Oil consumption in the largest consumer nation, the United States, dropped in 1982 by 3 million barrels per day. About half of this reduction was due to increased efficiency and fuel substitution. The remainder was the result of recession and unemployment.

THE THREE-PARTY CONFLICT

There is a three-party conflict among the oil-producing countries, oil companies, and consumers. Oil needs and problems are ubiquitous, permeating political realities and subject to economic constraints, resource allocations, and technological breakthroughs. They are further complicated by massive environmental impacts.

Today's energy crisis was triggered, but not caused, by the Arab-Israeli war of 1973 and the resulting oil embargo. The geometric growth in demand for energy among the developed nations had drawn far too little attention as the finite supplies were consumed.

The increasing demand for oil in all countries has caused it to become the most important energy source in the world. International oil is a multibillion-dollar industry that affects, in varying degrees, the balance of payments between oil-exporting and -consuming countries.

The cheapest and most abundant oil supplies are concentrated in the Middle East. A corollary to this oil concentration is the industrialized world's dependence on it. In 1973, when the Arabs implemented their threat to reduce production and imposed an embargo on oil supplies to the world, they also stepped up diplomatic efforts to have Israel withdraw from occupied Arab lands. Since then, the Middle East has continued to be a scene of strife among the world powers.

The conflicting, and in some ways contrasting, interests of the producing and consuming countries and multinational oil companies have gained the attention of the world's governments and people. The most powerful and least vulnerable among them, the United States, has become increasingly alarmed to find that it was importing ever-larger quantities of oil, mostly from the Middle East. Japan depends on the Middle East for about 50 percent of its oil. Western Europe is dependent to a lesser extent. This dependency makes Middle Eastern oil a vital necessity for the continued progress and survival of the industrialized world.

During the oil embargo, the Soviet Union sought to gain from the weakening of Western economies and the strains on the capitalist world. The Soviets maintained a public posture as friend of the Arabs in their war efforts against Israel, and a benefactor in the Arab nations' struggle against the Western oil monopolies. The Soviets have supported oil price increases. The emerging influence of the Soviet Union in the oil-producing and -exporting nations and the marked decline of U.S. influence in most of those countries is now threatening the world energy supply.

Projections as to the Soviet Union's oil prospects fluctuates. On the whole, it seems that the Soviet Union will continue to be one of the world's largest producers of oil and an exporter for the foreseeable future. Soviet influence in the Arab world, especially in Algeria, Iraq, Libya, Syria, and South Yemen, is strong. Revolution remains a possibility in conservative Arab states, and the Soviet Union can be expected to exploit any instability that may develop.

In a world where the price and availability of oil are critically important to many countries, the emergence of China as an oil-producing and -exporting country may be a welcome relief. Although the enthusiastic and rosy estimates of record-breaking production and new oil discoveries on land and offshore are encouraging, they are unrealistic. Those familiar with Chinese estimates and statistics on oil reserves and production know they are nothing more than speculation. At present China and the Soviet Union contribute little to the world oil trade. But the Soviets are capable of supplying much more than they do.

The OAPEC decision to impose a destination embargo in 1973 catapulted its parent organization, OPEC, into a position of great world power. OPEC suddenly found itself able to command oil prices higher than anyone had ever thought possible. Both Arab and non-Arab members of OPEC supported the price revolution as a political weapon against the United States and other supporters of Israel. But even after the embargo ended, over a decade ago, production and pricing decisions remained with OPEC.

There have been three oil regimes in world politics. The first did not reconcile the oil interests of consumers and producers, but it secured the interests of industrialized nations well enough to provide a solid base for mutual beneficence with remarkable longevity. The second regime gave the world low prices and increasing supplies. The third regime began with confrontation and ballooning oil prices. The confrontation is likely to continue until diversified energy sources are developed to replace oil as a vital commodity.

The third oil regime is dominated by three nations—the United States as the largest consumer, Saudi Arabia as the largest noncommunist producer, and the Soviet Union as the largest producer in the world, though mostly for domestic use. In the event of a new oil crisis, the Soviet Union may take advantage of its oil by selling more to needy countries. The Soviets may also be willing to sell or exchange oil for U.S. food and technology. Another new development that OPEC's price increases has precipitated is the effort to increase oil-production by non-OPEC nations. A significant development has been the growing importance of Mexico as a producer of oil and gas for the world market, particularly the U.S. market. Increased Mexican oil sales to the United States have lessened the latter's dependence on OPEC.

SUPPLY, DEMAND, AND PRICE

Perceptions of oil supply, demand, and price change continually. The 1973–74 Arab oil embargo, the 1978–79 interruption of Iranian exports, and the outbreak of war between Iraq and Iran are all periods in which the inventory levels are cited to explain why oil shortages are more serious than might have been expected.

A new outlook seems to be emerging in the 1980s. Many observers believe that, barring a major accident, the oil supply and demand balance will remain slack. World demand for oil is shrinking. To be sure, oil is still king, but the nature of its power is changing. Oil has ceased to rule through the pressure of its quantity. It now rules through price hikes. High prices have forced everybody to conserve and find alternatives to oil.

Three factors must not be overlooked in understanding the politics of oil supply, demand, and price. These are location, production, and consumption. In order to better understand the intricacies of the relationship of these factors and of the bargaining process that has evolved between producers and consumers, some analysis of the present energy situation is in order.

The expansion of the oil industry in the nineteenth century and the early years of the twentieth century has a sound basis in peace and in the efforts of many people to raise their standard of living. The growing indispensability of oil has made its availability imperative. Mounting demand makes its production and distribution attractive commercially. Never has oil supply and demand been more competitive than now.

It is important to note that oil is located randomly according to geological laws, with no regard for political boundaries or economics. It is found where nature put it. The need and demand for fossil fuels to furnish energy is a relatively new phenomenon. Only with the development of the modern industrial state have fossil fuels been used intensively for a variety of purposes.

The world shifted to oil during World War I to fuel its industrial, military, and technological machinery. It then became inevitable that the Middle East would be of central significance in the world power equation. For the first 70 years of this century, the potentials of its resources were not realized. Lacking indigenous technological knowledge, economically feeble, and subjected to the control of Western colonial powers, the oil-producing countries lacked the knowledge to determine the fate of their own resources. The key decisions concerning exploration, production, marketing, and pricing of petroleum products were made by large Western oil companies or their governments in Europe and the United States.

Following World War II and coinciding with the strong nationalist movements that swept the developing Third World, the oil-producing countries began to demand greater control over their resources. At first, these demands took the form of insistence on higher prices and larger royalties for oil. In short, the political leaders of the oil-rich countries fought, in the beginning, for better bargains. With the formation of OPEC in 1960, followed by the succession of meetings in Tripoli and Tehran in the early 1970s, and the quadrupling of prices during the Arab-Israeli war of October 1973, the entire process of international decision making about oil was revolutionized. The producing countries, now no longer merely content to drive better bargains, challenged the entire bargaining system and succeeded in wrestling complete control from the transnational oil organizations. From then on, the producing countries became the repositories of authority concerning the entire hydrocarbon production and exploitation industry. It is now the oil companies and the consuming countries which have been put on the defensive. In fact, the international oil companies are now, by and large, agents or contractors for the OPEC nations.

To further comprehend the problem of oil supply, demand, and price on a global basis, the balance between production and consumption in various countries and regions must be considered. The countries that produce more than they consume supply countries that produce less than they consume.

In the 1960s, some producing countries, notably the United States and the Soviet Union, were both exporters and importers of oil. During this period when U.S. production exceeded domestic consumption, the Caribbean region was a major supplier of Eastern-Hemisphere requirements. Most Caribbean oil came from Venezuela, which was second only to the United States in annual production.

Although its production has been rising, the Western Hemisphere does not produce enough oil to meet its own requirements.

Beginning in 1970, production was not increased to meet the ever-growing demand. Therefore, U.S. oil reserves began to deplete sharply. Although the demand for oil was growing in the United States, no important discoveries had been made. In this OPEC saw an opportunity to profit. In December 1970 OPEC called for, among other issues, an increase in the tax rate and the price of oil. Alarmed by the decline in domestic oil production, the U.S. Department of State called a meeting of the representatives of oil-importing countries in Paris to convince them that it was in their best interest to agree to the higher price of oil.

In the Tehran Agreement of February 1971 and the Tripoli Agreement of April 1971, OPEC fixed prices for oil at the production stage and provided for a 2.5 percent annual increase in price plus five cents per barrel to take into account the rate of inflation. These agreements were welcomed by the U.S. Department of State, which said that "the international oil business was entering an era of good feeling, one of stability that would last at least five years."[23]

The United States demonstrated its weakness as the largest consumer of oil by welcoming OPEC's decision to increase the price. Once this was done almost nothing could prevent subsequent price hikes. Iran, because of its credibility in the United States as a strong pro-Western nation and continuous supplier of oil to the West and Israel, could take the initiative in the Tehran and Tripoli agreements. "It was Iran that took the lead in asking for higher tax rates immediately after the early Libyan agreements, and Iran was the first to get a price increase for heavy crude oil."[24]

After slow growth in 1971, crude-oil production received new impetus in 1972 when production in all countries reached about 52 million b/d, or about 5 percent more than the year before. The year's growth was concentrated in a relatively small number of countries, foremost among which were Saudi Arabia and Iran.

On one hand Iran increased its oil production to take advantage of the higher price; on the other it took the lead in the Geneva Agreement of January 1972 to raise the price of oil by 20 cents per barrel to offset the U.S. dollar devaluation of August 1971. After that, OPEC, at a meeting in June 1973, raised the price of oil by another 15 cents per barrel, to offset another U.S. dollar devaluation in February 1973. In August, Iran wiped out the 19-year-old consortium agreement covering onshore oil production. Forthcoming was still another oil price hike. In September, OPEC decided to considerably increase posted prices, and thence royalties, effective October 16, 1973. The OPEC nations also "called for a renegotiation of the Tehran Agreement, noting that the current rate of inflation far exceeded the 2.5 percent rate contemplated in that agreement."[25] Both OPEC and the oil companies agreed to negotiate, but the negotiations were never completed. In October, OPEC members in the Persian Gulf, led by Iran, unilaterally increased the price of oil, raising it from $3.01 per barrel to $5.11. "For the first time, the oil-exporting nations took the pricing function completely into their own hands."[26] Thus, the established process of negotiation between OPEC and the oil companies ended abruptly. It was an overt political act on the part of OPEC, not an economic matter, to end negotiations prematurely in violation of the Tehran Agreement.

A FUNDAMENTAL CHANGE IN THE STRUCTURE

A fundamental change in the structure of the oil industry took place after OPEC reneged on the principle of the Tehran Agreement, which was supposed to last until 1976. Now the producing countries could dictate pricing and other conditions of oil sales. In this, Iran has been in the forefront, rather at the expense of Arab nations. The Iranian chairman of the gulf "chapter" of OPEC was present at the historic meeting of OAPEC in Kuwait on October 17, 1973.[27] The very fact that Iran, a non-Arab nation, was present at the meeting of OAPEC, an exclusive Arab oil cartel, was suspicious. As subsequent events showed, Iran had an ulterior motive.

At the beginning of the OAPEC meeting in Kuwait, Saudi Arabia successfully resisted attempts to set an embargo against the West. Even Kuwait announced that it would contribute $350 million to the warring Arab states' efforts but that it did not want to stop oil output.[28] It seems that the Iranian presence at the OAPEC meeting was to convince Arab nations to cut off oil supplies to pro-Israeli nations. The decision to cut off oil production was announced by the Iranian chairman before the handwritten Arabic text of the Arab oil ministers' resolution was distributed to reporters.[29] To maintain its posture of friendship toward the West, at the expense of the Arab nations, Iran wanted to increase oil production and sell it at an exorbitant price to the industrialized world, and even to Israel. Taking advantage of the embargo, Iran proposed increasing the posted price of oil to $23 per barrel at a meeting of OPEC in Tehran in December 1973.[30] At the instruction of then King Faisal, Saudi Arabian oil minister Sheikh Yamani not only resisted the move but clashed with the shah of Iran, who finally settled on $11.65 per barrel, effective January 1, 1974. This was the officially established price, but as happens in crises, Iran sold oil at more than $16 per barrel at auction while the embargo was in effect.[31]

The energy shortage became a major problem in many parts of the world during 1973–74. Especially hard hit were the industrialized nations of the West and Japan. The problem of fuel supply was caused by the continuing rapid growth of demand for energy, particularly in the form of petroleum products. For example, oil consumption in Western Europe rose sevenfold from 1956 to 1973, when the Arab countries of the Middle East and North Africa cut back production in retaliation for Western, principally U.S., support of Israel during the Arab-Israeli war. The Arab countries virtually declared an "oil war" on the West and Japan. The seriousness of the energy supply shortage was underscored in 1973 by Western Europe's dependence on the Middle East for more than 70 percent of its oil. Japan relied on the region for over 80 percent of its supply. Although the United States was much less dependent on imported oil, the Arab oil embargo had a considerable impact on the United States.

Whereas 1973 and 1974 had been years of confrontation in oil affairs, 1975 and 1976 were years of dialogue and discussion. Although a freeze on oil prices was agreed upon by OPEC in December 1974, a 10 percent increase was announced in September 1975. After maintaining Saudi Arabia's imposed price freeze for most of 1976, OPEC members split in December, with Saudi Arabia and the UAE raising oil prices 5 percent. The remaining members raised prices by 10 percent. The price hike on June 28, 1977 was between 16 and 24 percent. In December 1978, a further price increase of 14.5 percent was announced.

After the most fractious meeting in OPEC's history, in December 1979, the cartel failed to agree on any uniform price. As a result there was no official oil price for some time. Then a temporary compromise was reached in Vienna in September 1980. Under this agreement the price of Saudi oil would rise $2 per barrel to $30 while other OPEC members would freeze prices at existing levels, which averaged about $32 per barrel. In October 1981 OPEC fixed a new united price of $34 per barrel and froze it through the end of 1982. Saudi Arabia meticulously adhered to this price. The other OPEC members upped their benchmark price to $36.

REDISTRIBUTION OF WEALTH

The most obvious and immediate effect of the oil price hikes in the 1970s was a redistribution of wealth. While many nations sank into recession, OPEC countries suddenly became flush with money. By the end of 1980 they had accumulated $300 billion in foreign assets. The trade surplus of OPEC countries reached $152.5 billion.

The OPEC nations recycled their foreign exchange surpluses in three ways: (1) by buying Western consumer goods, military hardware, industrial equipment, food, and other commodities; (2) by investing their funds in development projects at home and abroad, especially in the United States and Western Europe; and (3) by lending money through official and private channels to oilless nations.

Higher oil prices have a relatively minor impact upon industrialized countries, since those countries can borrow to finance their imports until broader adjustment ensues. The nonindustrialized nations have no such options, and their economic hardship is not considered fundamental to global economic order. As Massachusetts Institute of Technology (MIT) economist Morris A. Adelman points out, "it's the underdeveloped countries that certainly are hurt as much or more than anyone by high oil prices, but they don't say a word."[32] Sooner or later underdeveloped nations are likely to insist on redress for their grievances, if the hopeless situation persists. Realizing this, oil exporters have begun to soften their attitude toward poor nations and are volunteering economic assistance, in some cases in greater amounts than Western nations had provided.

The escalation of oil prices has had a dramatic effect on international trade balances and monetary debt. Higher oil prices have caused trade deficits in almost all oil-importing countries. While wealthy nations with economic flexibility have been able to adjust to the shock of the oil-price revolution, developing countries are being overwhelmed with debt. By one account the total debt of developing countries has soared from $100 billion in 1973 to $500 billion at the end of 1981.[33]

To cope with this situation, the IMF established an oil-lending or recycling facility with OPEC-provided funds. Initially, OPEC contributed $9.6 billion for this purpose. Nearly all of this amount was lent to developing nations. The developed countries created within the IMF a $25 billion "safety net" fund for the 24 nations belonging to the Organization for Economic Cooperation and Development (OECD).

The higher prices have remained in effect for nearly a decade. Of particular importance are massive balance-of-payments problems and inflation, both of which continue to plague the world economy. The economic situation would imply that importing nations would suffer a corresponding balance-of-payments deficit. This deficit, due to oil purchases by developed nations, is estimated at $50 billion a year; that of developing nations is nearly $70 billion. The developed economies can somehow absorb this deficit, but the less developed economies just go deeper in debt. Many oilless countries remain permanently on the verge of bankruptcy.

With the transfer of wealth has come a concomitant shift in power. The major oil-producing states not only controlled a vital resource, without which industrialized economies would collapse, they also had accumulated huge financial reserves and assets. Although the petrodollars are invested in industrialized economies, they could be withdrawn by the OPEC nations to destabilize various currencies or weaken the economies of industrialized nations.

The damages that higher oil prices have inflicted on Western and Japanese economies handicap them in the arms race with the Soviet Union. This handicap works against the interests of oil rulers such as the Saudi monarchy, which depends on Western protection against communism and subversion. The discontent fostered by recession, inflation, and unemployment may feed authoritarianism in industrial nations. In the oilless countries of the Third World, governments struggling against worsening situations are growing more authoritarian, dictatorial, and repressive. Their leaders find aggressive foreign policy a means of diverting blame for their internal problems to foreign scapegoats. Rising tension is the outcome of such policies.

Third World countries were hurt more than industrialized nations by the oil embargo and the accompanying oil price increase. Fortunately, two developments have come to their rescue. First, OPEC has given a substantial share of its oil windfall to the Third World to support their balance of payments, help finance projects, and assist in creating jobs. Second, OPEC is also providing financial assistance for energy and food production in the Third World. This is helping some Third World countries in oil exploration.

Still, a tremendous economic disparity between oil haves and have-nots remains. Two kinds of interlocking diplomacy are emerging: resource diplomacy and basket diplomacy. Resource diplomacy is evolving as a result of seemingly limitless demand on critical resources. Basket diplomacy is an attempt to turn begging hands into productive hands by those nations which are victims of resource constraints. These nations are relying to a greater degree on foreign aid.

One optimistic sign for outsiders is that after more than a decade of unchallenged control over steadily rising oil prices, dissension and discontent have broken out within the 13 members of OPEC. At issue is how to deal with the precipitous drop in prices and worldwide demand. As a result of recession in the United States and conservation measures practiced by all oil-consuming nations, worldwide production of oil has exceeded demand. The oil crisis has brought about an oil glut, an oversupply. This, coupled with the psychological insularity of the OPEC nations, has already produced the desired effect on the price of oil.

To sustain higher oil prices OPEC nations are trying to cut back production. Not all of them can. Some, such as Algeria, Iraq, and Indonesia, must increase their production, not only to keep up with their development plans and other programs, but also to save themselves from bankruptcy. Can the largest OPEC producer, Saudi Arabia, lower its production of 4.5 million b/d enough to stop the glut? The Saudi ruling family has so many financial obligations as a result of its one-resource economy that it cannot reduce its income below that represented possibly by 4 million b/d. Saudi Arabia is feeling a financial pinch. After dropping production last year, Saudi Arabia has been using interest earned on its $175 billion investments, instead of simply reinvesting the money. There are indications that the Saudis may dip into principal.[34] Therefore, to reduce oil production below the current level would create internal and external political and economic trouble that the Saudi rulers are anxious to avoid. After all, they do not want to see Saudi Arabia becoming the next Iran. On top of that, supplies from non-OPEC nations climbed above OPEC's for the first time last year. For obvious reasons, OPEC nations do not want to lose their market to the non-OPEC producers.

Uncertainties will continue concerning both the demand for oil and terms on which oil will be supplied in the future. Demand will depend on the growth rate of the world economy and the reactions to high prices.

As production capacity increases all over the world and demand decreases, we can anticipate lower prices. The drive toward world energy self-sufficiency, particularly in the United States, through conservation, recovery of onshore and offshore oil, and development of alternative energy sources, will go a long way in reducing U.S. dependency on imported oil.

In the meantime, if another Arab-Israeli war were to break out, the Arabs could be expected to use their oil weapon again. In theory, the Arabs can be expected to use their oil for political advantage; but in practice, another embargo will probably have more dangerous consequences for the Arabs, as the non-OPEC nations, including the Soviet Union, have sufficient oil to take over the world market. A selective embargo against the United States would stand little chance of success. However, a total embargo against the entire noncommunist world might be met by a military response from the West. Therefore, Arab oil as a political weapon has only a minimal prospect of success.

In sum, another oil embargo like the one in 1973 would be economically disastrous for the oil-producing and exporting countries. The Soviet Union could sell more oil to the industrialized capitalist world than Saudi Arabia. The OPEC countries can no longer afford to raise oil prices indiscriminately. They would be hurting more and more their friends, the oilless Third World countries. Oil will play an insignificant role in the Arab attempt to settle differences with Israel. The unavailability of oil as an economic weapon may tend to encourage highjacking and hostage situations. Finally, oil prices will be subject to manipulation by OPEC, but prices may not rise spectacularly nor drop significantly.

If the glut continues, the present fragile market structure may offer° some stability for consumers. But if supplies again become tight, consumers are likely to be faced with renewed instability in the world oil market. It seems prudent that consumers reduce their dependence on imported oil.

NOTES

1. The United States comprises about 6 percent of the total world population, but it uses more than 30 percent of the world's energy.

2. See the full text of the final communiqué issued at the conclusion of the seventh annual economic summit attended by heads of government from the United States, Canada, Japan, France, West Germany, Italy, and the United Kingdom, in Chau T. Phan, ed., *World Politics 82/83*, Annual Edition (1982): 76–78.

3. Linda B. Miller, "Energy and Alliance Politics: Lessons of a Decade," *World Today* 39, no. 12 (December 1983): 481.

4. *Washington Post*, February 14, 1974.

5. Ibid., November 16, 1974.

6. *Newsweek*, December 23, 1974.

7. United States Department of State, *Energy: Continuing Crisis* (Washington, D.C.: November 18, 1980).

8. Ibid.

9. United States Department of Energy, *The National Energy Policy Plan* (Washington, D.C.: October 1983).

10. United States Department of Energy, *Secretary's Report to Congress* (Washington, D.C.: September 1983), p. 84.

11. Melvin R. Laird, "Let's Meet the Energy Crunch Now," *Reader's Digest* 106, no. 633 (January 1976): 50.

12. *The World Almanac* (New York: Newspaper Enterprise Association, 1975), Annual Edition, p. 114.

13. David J. Rose, "Energy Policy in the U.S.," *Scientific American* 230, no. 1 (January 1974), p. 359.

14. *Newsweek*, November 12, 1973.

15. MIT Energy Laboratory Policy Study Group, *Energy Self-Sufficiency: An Economic Evaluation* (Washington, D.C.: American Enterprise Institute for Public Policy Research, 1974), pp. 52–53.

16. National Petroleum Council, *U.S. Energy Outlook: An Interim Report* (Washington, D.C.: National Petroleum Council, 1972), p. 20; *Newsweek*, December 10, 1973.

17. Lawrence Rocks and Richard P. Runyon, *The Energy Crisis* (New York: Crown Publishers, 1972), p. 40.

18. Ibid., p. 39.

19. Allen L. Hammond, William D. Metz, and Thomas H. Maugh, *Energy and the Future* (Washington, D.C.: American Association for the Advancement of Science, 1973), p. 12.

20. Wilson Clark, *Energy for Survival: The Alternative to Extinction* (Garden City, N.Y.: Anchor Books, 1974), p. 370.

21. Hammond et al., *Energy and the Future*, p. 12.

22. The seven sisters are Exxon, Mobil, Socal, Texaco, Gulf, British Petroleum, and Royal Dutch Shell.

23. Roger LeRoy Miller, *The Economics of Energy: What Went Wrong and How We Can Fix It* (New York: William Morrow, 1974), pp. 28–29.

24. Taki Rifai, *The Pricing of Crude Oil: Economic and Strategic Guidelines for an International Energy Policy* (New York: Praeger, 1974), p. 263.

25. Neil H. Jacoby, *Multinational Oil: A Study in Industrial Dynamics* (New York: Macmillan, 1974), p. 259.

26. Ibid.

27. Insight Team of the Sunday Times, *Insight on the Middle East War* (London: Andre Deutsch, 1974), pp. 178–79.

28. *New York Times*, October 18, 1973.

29. Insight Team of the Sunday Times, loc. cit.

30. *New York Times Magazine*, March 24, 1974, p. 13.

31. Ralph H. Magnus, "Middle East Oil," *Current History* 68, no. 402 (February 1975): 50.

32. Miller, *The Economics of Energy*, p. 30.

33. *New York Times*, August 29, 1982.

34. Ibid., November 28, 1982.

3

The Middle East Oil:
New Wealth, New Power

The increasing demand for oil in both developed and developing countries has caused oil to become the most important source of energy in the world. Oil is a multibillion-dollar international industry that affects, in varying deg. es, the balance of payments of oil-exporting and -importing countries. Because of the Arab oil embargo, oil has been enmeshed into international political and economic relations.

The Middle East is, by a wide margin, the main source of oil to the industrialized nations of Western Europe and Japan. The political potential in reserves of such a vital commodity is obvious. Oil has become the lifeblood of industrialized society, a resource which nations, especially the technologically advanced ones, cannot ignore. In addition to its economic value, oil is also an important strategic commodity, vital not only to industry but to national security.

As discussed in Chapter One, Middle East oil is Arab oil except that produced by Iran. These Arab states control about 34 percent of global production and 52 percent of oil reserves (see Tables 3.1 and 3.2). Moreover, supply is inelastic, so a loss of oil in one area cannot be quickly made up by increasing production in other regions.

Although the United States is much less dependent on Arab oil than its European and Japanese allies, Western economic and national security is quite dependent on the maintenance of the flow of oil. This leads us to assume that the Arabs possess a weapon potentially

TABLE 3.1. Middle East Oil Discoveries: Cumulative Production and Remaining Reserves (Billions of Barrels)

Country	Cumulative Production as of January 1, 1984	Estimated Proved Reserves as of January 1				Total Discoveries (Reserves + Cum. Prodn.) as of January 1, 1984
		1950	1965	1980	1984	
Bahrain	0.7	0.3	0.3	0.2	0.2	0.9
Iran	31.8	13.0	38.0	58.0	51.0	82.8
Iraq	16.6	8.7	25.0	31.0	43.0	59.6
Kuwait*	23.0	15.0	69.3	68.5	66.8	89.8
Oman	1.9	–	0.5	2.4	2.8	4.7
Qatar	3.6	1.0	3.5	3.8	3.3	6.9
Saudi Arabia*	50.1	10.0	66.8	166.5	168.9	219.0
United Arab Emirates	8.6	–	7.7	29.4	32.3	40.9
Total Gulf Area	136.3	48.0	211.1	359.8	368.3	504.6
Egypt	3.0	0.2	1.5	3.1	3.5	6.5
Total Middle East	139.3	48.2	212.6	362.9	371.8	511.1
United States	136.9	26.2	34.5	26.5	27.3	164.2
Total World	515.0	95.0	341.3	641.6	669.3	1,184.3

*Includes one-half of Neutral Zone.

Sources: Oil & Gas Journal, World Oil, Exxon Corporation estimates, and Exxon Background Series, *Middle East Oil and Gas* (New York, December 1984). Reprinted with permission of Exxon Corporation, 1251 Avenue of the Americas, New York, N.Y. 10020.

TABLE 3.2. Middle East Crude Oil Production

| | Thousands of Barrels per Day | | | | | | 1983 | |
Country	1950	1965	1975	1979	1980	1983	Producing Oil Wells	Daily Average Bbl/Well
Bahrain	30	57	57	50	48	41	243	169
Iran	664	1,886	5,350	3,168	1,662	2,492	530	4,702
Iraq	136	1,322	2,262	3,477	2,514	922	290	3,179
Kuwait*	344	2,351	2,087	2,497	1,661	1,076	755	1,425
Oman	–	–	341	295	282	376	477	788
Qatar	34	231	437	508	471	295	125	2,360
Saudi Arabia*	547	2,206	7,075	9,535	9,903	5,062	780	6,490
United Arab Emirates	–	282	1,694	1,831	1,702	1,119	399	2,805
Total Gulf Area	1,755	8,335	19,303	21,361	18,243	11,383	3,599	3,163
Egypt	45	125	231	525	595	689	500	1,378
Total Middle East	1,800	8,460	19,534	21,886	18,838	12,072	4,099	2,945
United States	5,407	7,804	8,362	8,533	8,597	8,680	636,900	14
World	10,428	30,308	53,418	62,658	59,464	52,621	NA	NA

*Includes one-half of Neutral Zone.

NA = Not Available.

Sources: International Petroleum Annual, U.S. Bureau of Mines; International Energy Statistical Review, U.S. Central Intelligence Agency; Oil & Gas Journal, and Exxon Background Series, Middle East Oil and Gas (New York, December 1984). Reprinted with permission of Exxon Corporation, 1251 Avenue of the Americas, New York, N.Y. 10020.

capable of providing their foreign policy with leverage, which can be used to influence the course of world events, especially Middle Eastern issues.

The U.S. position in the Middle East has been challenged by the rising power of the Soviet Union and the growing independence of the Arab nations. While this rivalry was taking place between the superpowers, a new power center emerged: the Arab oil-producing countries had gathered and formed OAPEC in 1967 to reduce the power of oil companies and control their own oil resources.

SIGNIFICANCE OF ARAB OIL POWER

Saudi Arabia is the largest oil exporter in the world. Its share of the world energy and oil market has increased threefold during the last two decades. In 1981, Saudi Arabia supplied over 15 percent of the world's oil and 7.4 percent of all energy consumed. The revenue earned by Saudi Arabia increased from $334 million in 1960 to a record $101 billion in 1981.[1]

The Arab motivation for using oil as a political weapon was: (1) to help speed Israeli withdrawal from Arab occupied lands; (2) to develop a diverse economic base through cooperation with industrialized nations; and (3) to assume a leadership role in the world.

The Arab decision to use oil politically was made on October 17, 1973. The OAPEC oil ministers, meeting in Kuwait, opted for a discriminatory oil export policy which had two objectives: to maintain their export to states sympathetic with Arab political goals and punish unfriendly states by gradually depriving them of their normal oil shipments. The primary target of this policy was the United States. Arab policymakers seem to have believed that a restrictive oil policy against the United States, at a time when Middle East oil was considered necessary, would eventually coerce the United States to reconsider its strong support for Israel. It was believed that U.S. influence would compel Israel to evacuate all of the occupied Arab territories and restore the legitimate rights of the Palestinian people.

It seems clear that the Arab oil embargo was an application of the idea that economic power of nations is an integral part of their foreign policy. If we superficially examine the apparent objectives of the Arab oil policy, it could be considered a success. However, we should not assume that the use of economic power to achieve political goals guarantees success. While the possession or control of a valuable resource such as oil theoretically provides a nation or group of nations with the capacity to influence others, effective coercion is determined by the validity of the assumption on which the decision to embark on a campaign of economic warfare is based.

It is difficult to find in Arab oil anything of "Arabness" except the name. What we call "Arab" oil is rather Saudi, Kuwaiti, Iraqi, Algerian, or Libyan oil. In other words, when it comes to oil, local identity prevails over Arab unity. This is better reflected in the social and economic situation of the different countries belonging to the Arab family. For instance, while the UAE enjoys an average income of no less than $22,870, South Yemen's per capita income is less than $200. Further, if the Arab Middle East is viewed as a unit, per-capita income is less than $800, hardly comparable to Southern Europe or even Israel, let alone the United States, Canada, or Western Europe.[2]

What is true for any assumed pan-Arab oil characteristics is true for the wealth derived from it. In this vein, it is instructive to note that Saudi Arabia, Kuwait, Qatar, and the UAE (which make up less than 8 percent of the Arab population) collect no less than 70 percent of oil revenue. The majority of this revenue is not invested in the Arab world, but in Western markets. While the Arabs appear to have a tremendous reserve of hard currency, many Arab states continue to receive foreign aid and are in urgent need of foreign exchange.

Thus, the apparently huge surplus of petrodollars does not greatly benefit the Arab world as a whole. Only a few of the major producers benefit. Even oil-producing states like Libya, Iraq, and Algeria do not enjoy large reserves of foreign exchange.

Saudi Arabia's huge reserves are less a reflection of great national wealth than of an extremely limited capacity to absorb its petrodollars. This limited capacity is a reflection of what remains a backward economy. The lack of economic development in producing countries prior to 1973–74 can be attributed to several factors. One was the historically low price of oil on the world market. Prior to the rapid price increases following the Arab oil embargo, a poor educational base, a lack of economic foresight, disinterested planning, a lack of national resources other than oil, poor spending patterns, and an overall lack of investment capital combined to produce extremely limited economic development. Thus, the national wealth that foreign exchange surpluses seem to indicate is largely illusory. As stated above, Arab petrodollar reserves are a reflection of the area's lack of development rather than an arbitrary and unrealistic oil price system.

The Arab people are not benefiting from OAPEC earnings to a great extent. The importance of petrodollars in inter-Arab politics is significant. Hans Morgenthau maintained that "international politics, like all politics, is a struggle for power."[3] Further, Morgenthau stated that there are several components which combine to generate power, not the least of which is economic or financial policy. Accordingly, economic and financial policy in international relations "must be judged primarily from the point of view of their contribution to national power."[4] If Morgenthau is correct and his perspective valid, then it is difficult to escape the conclusion that Arab wealth, particularly Saudi Arabia's, has provided some states with an extraordinary degree of power in inter-Arab politics. In fact, Saudi Arabia's financial power has tilted the balance between conservative and radical states in favor of the conservatives.

When Egypt's President Nasser assumed power in 1956, the pan-Arab movement was revived. Consequently, a rivalry developed between the Egyptian-led radical bloc and the conservative bloc led by Saudi Arabia. By 1958, several developments seemed to indicate that the radical movement would eventually prevail. Egypt and Syria were united in 1958, the Iraqi monarchy had been overthrown, and there was civil unrest in Jordan and a civil war in Lebanon. Only with British intervention in Jordan and U.S. intervention in Lebanon could the conservative regimes maintain control and curtail the radicals.

Although the Egyptian-Syrian union was destroyed in 1961 by a Syrian military coup, Egypt retained its Arab socialist ideology as the basis of its foreign and domestic policies. Cairo continued as the meeting place for Arab revolutionaries since Egypt maintained its ideological leadership in the Arab world. The major tenet of Arab socialism preached by Nasser and broadcast by the "Voice of the Arabs" was the removal of all foreign control over the Arab world.

While Arab revolutionaries looked to Cairo as their ideological homeland and capital, conservatives perceived Egypt as an evil power subsidizing the forces of unrest and political instability throughout the Arab world. In 1962, when Nasser dispatched Egyptian troops to Yemen to aid revolutionary forces, conservatives feared the Egyptian expedition would hasten a new era of Arab radicalism.

According to Malcolm H. Kerr, Nasser "had seized on the revolution in Yemen in 1962 as an opportunity . . . to regain the initiative in Arab affairs on the basis of revolutionary leadership."[5] In response, the conservatives seized upon Yemen as a means to tie Nasser's hands, then defeat him and strengthen their ideology. In short, Yemen became the arena for the battle for ascendancy in Arab politics.

For Nasser, success in Yemen meant gaining an ideological foothold on the traditionally conservative Arabian Peninsula, an event which would significantly advance the Arab revolutionary cause. Accordingly, the leader of the conservative bloc, Saudi Arabia, sought with all its resources to deny Nasser success. Because Saudi Arabia's limited military capability made direct intervention impossible, King Faisal was compelled to wage an indirect and cheap war for the forces of conservatism. Saudi Arabia committed money and military hardware to the Yemeni royalists, while the latter fought alone on behalf of the conservatives.

The eventual victory of the conservatives in Yemen was not really a military defeat of the Egyptians, but the result of a standoff. Because Nasser could not produce a definitive Republican success, given the time and material he had committed, it was necessary for him to learn to live with Saudi Arabia. The agreement which Nasser and Faisal reached in Jedda on August 24, 1965, may be said to have been the first of a series of compromises which would eventually remove Egypt from the revolutionary bloc and pave the way for Saudi Arabia's ascendancy in the Arab world.

Given the outcome of the Yemeni civil war, Egypt's devastating defeat in the 1967 Arab-Israeli War and its subsequent economic dependence on the conservative bloc[6] effectively formalized the de-radicalization of the Egyptian regime's Arab policy. More important, Egypt's defeat signaled the ascendancy of the conservatives. In the months following the June war, Egypt's role in Arab politics was altered significantly. The leader of the radical bloc, Egypt, became the mediator between conservative and revolutionary states. This trend was a result of the growing rift between Egypt on one hand, and Syria and Iraq on the other, in the aftermath of the war. Both Syria and Iraq not only refused to recognize United Nations Resolution 242, but also condemned it. Egypt and Jordan, by comparison, took a giant step away from the revolutionary bloc by accepting Resolution 242, a decision which implicitly recognized Israel's right to exist as an independent and sovereign state. As a consequence, both revolutionary Egypt and reactionary Jordan coordinated their policies during Gunnar Jarring's Middle East mission.[7] Syria not only refused to cooperate with Jarring, but also refused to accept a conservative subsidy by not attending the Khartoum summit.

Egypt's departure from the Arab radicals' ranks was further confirmed by its reaction to the fighting between Palestinian guerrillas and the Jordanian army in September 1970. Not only did Egypt refuse to intervene on behalf of the Palestinian resistance forces, but it also failed to mediate in the dispute until the Jordanian army had beaten the Palestinian forces decisively. Egypt's role in Palestinian-Jordanian fighting contributed to a large extent to the end of the radical influence in Jordan.

In summary, there have been many incidents since Egypt's defeat in 1967 which reflect its de-radicalization. The turning point of Egypt's political shift was, of course, "when Saudi Arabia's income ... [became] essential to Egypt for its recovery from defeat."[8] Despite the Saudis' suspicion of Nasser, the traditional feuds between the monarchist Saudis and socialist Egypt were set aside. Conservative Egyptian compromises also reflected a trend in which Egypt and Saudi Arabia found it mutually beneficial to coordinate their policies.

THE RISING POWER OF SAUDI ARABIA

With Anwar Sadat's assumption of the Egyptian presidency, Egypt's need for conservative economic support and Sadat's need for Saudi political support became even greater. Sadat enjoyed neither the prestige nor the influence of Nasser. He lacked Nasser's charismatic leadership, a quality which had enabled Nasser to surmount political crises and to outlive his own mistakes. However, Sadat's beliefs about the Arab-Israeli conflict and the means of resolving it were similar to those of Saudi Arabia's King Faisal. Since both heads of state believed a U. S. role was crucial to ending the Middle East conflict, two factors combined to establish a loosely defined alliance.

First, Faisal needed Egypt's political dependence if Saudi policy was to gain ascendancy in the Arab world. Despite the failure of the union with Syria, the embarrassment in Yemen, the debacle in 1967, and Nasser's death, Egypt in 1970 was still the leader of Arab politics. Nasser's political legacy was very much alive. Second, Sadat needed Saudi financial support to fund his fledgling government's domestic programs. Thus, the political-economic interdependence allowed Saudi Arabia and Egypt to reach something of a "meeting of the minds."

The previous discussion is not meant to imply that the convergence of Egyptian and Saudi policy took place overnight. For one thing, Faisal continued to be suspicious of the Egyptian socialist tradition and the large Soviet presence, both of which were potential threats to Arab conservatism. Likewise, Sadat had to be cautious because of Nasser's legacy and Egypt's socialist identification. Thus, Sadat was compelled to pay lip service to the idea of continuity.

Accordingly, Sadat began his regime with an indirect and gradual disavowal of the Nasser legacy. In practice, Sadat's shift took the form of deemphasizing socialist ideology and pan-Arabism, lessening government control over economic activity, and opening Egypt's market to Western goods and investments. By July 1972, the time was ripe for Sadat to set out on a course dictated by the new political realities. Only one year after signing a Treaty of Friendship and Cooperation with the Soviet Union, Sadat expelled Soviet advisors and military personnel.[9] Simultaneously, Sadat imprisoned several prominent Nasserites (Ali Sabri, Sharaoui Gumea, and General Mohammed Fawzi) and dismissed several important ones (Haykal and Hussein al-Shafei). Following this purge, Sadat freed many of Nasser's political enemies, including General Mohammed Nagib and the twin brothers Ali and Mustapha Amin.[10]

In international relations, Sadat repudiated Nasser's legacy by abandoning leadership of the nonaligned Third World and militant support for the world's nationalist movements. Radical pan-Arabism was disavowed and its leaders, such as Libya's Colonel Muammar Qadhafi, became the targets of Egypt's hostile propaganda. Furthermore, the Soviet Union, formerly Egypt's and the radical bloc's staunchest ally, had been identified by Sadat and the Egyptian press as the principal enemy of the Arab cause. Conversely, from 1973 to the present, the United States has, in Egyptian eyes, become the only major power capable of enabling Egypt and the Arab world to achieve their international and domestic goals.

Given Sadat's dependence on Saudi Arabia's support for Egypt's post-1973 policies, it is conceivable that Egypt's radical foreign policy shift has resulted from direct Saudi influence. Moreover, Saudi Arabia's direct and aggressive intervention in the 1973 Arab-Israeli

War by taking the lead in the OAPEC embargo[11] may be seen as Saudi Arabia's initial move to wrest the leadership of the Arab world from Egypt.

Historically, Saudi Arabia's role in the Arab-Israeli conflict has been marginal. In the crucial period of political transaction preceding the 1967 war, King Faisal saw fit to make an official tour of Europe. Faisal had little influence in Arab affairs and very little interest in them. In the aftermath of the conflict, he imposed a shortlived embargo on oil shipments to both England and the United States, but as Nadav Safran has observed, "Saudi Arabia cut the flow of oil [in 1967] involuntarily under pressure by Nasser and therefore did not enforce the measure strictly and cancelled it as soon as possible."[12]

In contrast to its 1967 reaction,[13] Saudi Arabia took the lead in introducing the oil weapon in 1973. Saudi Arabia has surprisingly placed all of its prestige and financial resources behind resolving the Middle East conflict. Saudi diplomacy, previously known for its calmness and rationality, suddenly became almost bellicose, threatening the West with loss of crude supplies unless the Western allies pressured Israel to withdraw from the occupied territories. One could observe that Riyadh assumed, during the six months preceding the October war, much of the same character that had typified Cairo prior to the June 1967 war: the official and aggressive spokesman of the Arab cause. In retrospect, the positions taken by Egypt and Syria were secondary to those issued from the Saudi capital.

Faisal's leadership in employing the oil weapon may be viewed as a confirmation of Saudi Arabia's leadership of the Arab world. Before 1973, Saudi Arabia had strongly refused to honor the pleas of Nasser and other radical leaders to use oil as a political weapon. Faisal's opposition was viewed as a result of the conservative-radical split. Faisal "feared that once he had sprung the oil weapon others, particularly Nasser, might be able to arrogate the right to decide when and how it was to be used."[14]

In 1967, Faisal feared that his wholehearted support of the oil weapon would have been more beneficial politically to the radical cause. Because of a shift in the intra-Arab balance of power, in 1973 Faisal was able to retain absolute control over the embargo and harvest political advantages.[15] Saudi participation not only made the political use of oil a reality, but also brought the Saudis new popularity and greater influence. One might even conclude that the major reason Faisal chose to enter the Arab-Israeli conflict was to exploit the Arab world's dependence on Saudi participation in its oil diplomacy, and strengthen and perpetuate conservative influence.

THE AFFIRMATION OF SAUDI LEADERSHIP

Saudi leadership in Arab affairs has been established since the conclusion of the October war. Evidence of Saudi power and influence can be found in its economic assistance to those Arab governments which have committed themselves to negotiating peace with Israel. Further evidence of Saudi influence in settling the Middle East dispute was the special attention U.S. Secretary of State Henry Kissinger gave to Saudi Arabia during his shuttle diplomacy. Kissinger spent almost as much time in Riyadh as he did in Cairo, Amman, and Tel Aviv. In retrospect, it may have been Saudi pressure which induced Syria and Egypt to accept the Kissinger approach.

Edward F. Sheehan, a leading Middle East analyst, has gone so far as to identify Saudi Arabia as the major factor for the success of the U.S. peace effort. Sheehan called Saudi aid to Egypt and Syria the "financial cement" of Kissinger's diplomacy. "Indeed, it may not be too much to say that in large measure King Faisal is financing the method Secretary of State Kissinger has chosen to achieve peace."[16] Sheehan based his analysis on the fact that:

> the fundamental of King Faisal's policy is to prevent Soviet hegemony in the Arab world and to strengthen the forces of moderation as insurance for his own conservative monarchy. A new Middle Eastern war, he knows, will risk unleashing a radical recrudescence he cannot contain and thrusting the Arab states into a new dependency on Moscow that might menace the stability of his own kingdom.[17]

Given Faisal's power and influence in Arab affairs as well as the stake he had in limiting the Soviet presence in the Middle East, it is no small wonder that Kissinger and the leaders of the Arab confrontation states felt compelled to consult with the Saudi monarch. For instance, the rift that developed between Sadat and Hafez al-Assad after the signing of the second Sinai Agreement could not have been bridged without Saudi influence. The same could be said of the Saudi role in Lebanon, especially after Syria's direct intervention in the face of Egyptian and Palestinian opposition. The Saudis were successful in arranging a compromise in both cases. In exchange for Sadat's acquiescence to Syria's virtual occupation of Lebanon, Assad agreed not to oppose Sadat's peace initiative any longer.[18]

The Egyptian-Syrian compromise discussed above was the major outcome of the Saudi-sponsored Riyadh summit of October 1976. The Riyadh meeting was notable since it was the first time Sadat and Assad had met since late 1974. But the meeting also reflected Saudi influence and made possible a definitive cease-fire in the Lebanese civil war.

The ability of Saudi Arabia's King Khalid to achieve the compromise did not go unnoticed. The *New York Times*' Middle East correspondent, Henry Tanner, analyzed the conference's outcome succinctly: "The Saudi royal family imposed a cease-fire on the Syrians and the Palestinians."[19] Following the conclusion of the minisummit, Saudi Arabia summoned the chiefs of state of the Arab League to sanction the Lebanese cease-fire.[20] The Saudi-initiated meeting took place on October 25–26, 1976, in Cairo, where the 20 respective chiefs of state endorsed the Saudi compromise. Tanner's conclusion was also clear: "These days a joint [Arab] strategy, regardless of its authors, needs the stamp and seal of Saudi moral endorsement."[21] While no one can seriously discount Saudi claims on the moral leadership of the Arab world, Saudi influence rested on more than righteousness. Both Syria and Egypt were and continue to be dependent on Saudi financial aid. Saudi Arabia's apparently boundless wealth fuels Saudi influence in the Arab world. Saudi wealth and predominant share of OAPEC's proven reserves gives King Fahd control over any decision concerning the collective use of Arab oil.[22] If the share of OAPEC reserves controlled by Saudi Arabia's political allies is added to the Saudi share, the conservative bloc controls fully 75 percent of OAPEC's total reserves. In comparison, the radical oil producers—Iraq, Libya, and Algeria—control only 19 percent of Arab reserves. Production capacity also strengthens the conservative bloc's hand. From 1973 to 1975, the conservative bloc produced an average of 70.3 percent of OAPEC's total output. The Saudi share alone averaged 43.9 percent. Saudi Arabia is the only OAPEC member able to increase production significantly. At any time, the Saudi fields could increase production by 3.5 to 4.5 million b/d.[23] Out of OAPEC's additional production capacity of 7.5 million b/d, the radical states combined can produce only 2 million b/d, which obviously weakens their bargaining position.[24] In summary, it is ironic that while the radical states have consistently advocated the use of the oil weapon, it is the conservative states which control the issue.

OAPEC oil could never be used politically without the cooperation of the Saudi-led conservatives. Without Saudi consent, the oil weapon is quite like a large artillery piece without ammunition. Therefore, the conservative bloc can render impotent any maverick attempt by the radicals to impose an oil policy similar to that of 1973.[25]

The conservative OAPEC actually controls the use of oil for itself and OPEC. The conservative bloc's influence on the use of the oil weapon is manifest in its foreign currency reserves. The reserves allow OAPEC to cut production drastically for several months without significant loss or suffering. If the radicals attempted to reduce production significantly, the consequent impact on their domestic economies would likely lead to civil unrest and political instability.

In conclusion, the combination of factors such as oil production, reserves, surplus capacity, and surplus capital makes Saudi Arabia and its conservative allies the only states capable of determining when, whether, how, and for how long oil could benefit Arab foreign policy.

OIL AND SAUDI-UNITED STATES RELATIONS

Whether or not Saudi Arabia will support the political use of OAPEC's oil wealth may depend on its relations with the United States. Several aspects of current U.S.-Saudi Arabian relations and a brief review of the 1973–74 embargo suggest that Saudi Arabia's leaders may not opt to use oil as a political weapon again in the near future.

It must be remembered that in 1973–74 the OAPEC oil weapon was directed at the West to induce those countries to pressure Israel into giving up Arab territories occupied since 1967. The rationale for making the United States the prime target was that Israel was and continues to be overwhelmingly dependent on U.S. military and economic aid. Logically, the United States was the only state capable of forcing Israel to meet Arab demands.

While Saudi Arabia controls the OAPEC oil weapon, the Saudis are also extremely dependent on the United States as a source of military hardware and technical assistance.

Saudi Arabia also has been almost as friendly to the United States in terms of international politics as Israel. It is useful to review Saudi

relations with the communist world. While Israel had friendly relations with Eastern-bloc countries prior to 1967, Saudi Arabia has
consistently refused to have any relations with the Eastern bloc, the
Soviet Union, or China. The Saudi leadership historically has perceived
Israel's existence as an international communist conspiracy. In fact,
it is difficult to differentiate between the uses of Zionism, communism, and socialism in official Saudi rhetoric. According to Edward F.
Sheehan, one of the major reasons Saudi Arabia accepted United
Nations Resolution 242 was "to work for Israel's containment behind
her 1967 frontiers as the best means of keeping the region stable and
of reducing Soviet influence."[26]

Another indication of Saudi aversion to communism is a statement King Khalid made in mid-1975 expressing his concern over
seeing "billions of dollars of Arab money [flowing] to the Soviets
for arms." Khalid called upon the United States to arm Egypt and
Syria as a means of stopping the flow of money to Moscow.[27] One
analyst has noted that Saudi foreign policy "has been predicated
upon a staunch opposition to communism internationally and
regionally . . . [which has] . . . contributed to a foreign policy based
on a close relationship with the United States and other Western
powers."[28]

Given the affinity of U.S. and Saudi policy toward the communist world, it is not surprising that since the 1973–74 embargo the
United States and Saudi Arabia have created a web of unprecedented
mutual dependence. The cornerstone of this interdependency is found
officially in the set of broad economic and political cooperative
agreements that were signed on June 8, 1974. The innovative features
of these agreements were the establishment of a permanent Joint
Economic Commission and a Joint Cooperative Security Commission.
Both commissions are to meet regularly and be led by ministerial
level officials from each nation.

Commenting on the Cooperative Security Commission, Ronald
Koven wrote in the *Washington Post* that although "U.S. officials
stressed that this new form of military relationship does not involve
any American defensive commitment—it is apparently designed to
satisfy Saudi hunger for a U.S. moral commitment to support them
[Saudis] against potential aggressors."[29]

Within the framework of the 1974 U.S.-Saudi agreements—
termed a positive alliance—the United States is currently enjoying a

heretofore unknown influence on Saudi Arabia. Besides the large contingent of U.S. military advisors, reflecting the growing Saudi dependence on American military assistance and technology, the Saudis have gone a long way toward increasing their dependence in other fields. For instance, since 1974 Saudi Arabia has signed agreements with the following U.S. groups: the Labor Department, to develop Saudi manpower training programs; the National Science Foundation, to develop a Saudi science center; the Census Bureau, to develop a Saudi statistical base; the Agriculture Department and Interior Department, to aid Saudi farm and water resources; the Treasury Department, to plan and purchase equipment for a new Saudi electrification system and a host of other programs.[30]

Saudi dependence on the United States also extends to the Saudi presence in the United States. Huge investments in U.S. markets and banks made by the Saudis and their OAPEC allies have probably increased Saudi dependence on the United States. The more the Saudis invest in the United States, the more U.S.-Saudi interdependence increases. In the words of William Stoltzful, former U.S. ambassador to Kuwait: "The more investments and ventures of all kinds that they [Saudis] have in the United States or with U.S. companies in the Mideast, the more importance they will attach to U.S. strength and viability not only domestically but in their own countries as well."[31] He went so far as to urge U.S. leaders to facilitate Arab investments and encourage all kinds of Arab ownership in the United States, particularly in the field of oil. "The more ownership of downstream [oil] facilities they [Saudis and conservatives] have, the more vested interest they will have in both ends of the oil industry, and sellers and buyers."[32]

What is apparent from the previous discussion of U.S.-Saudi relations is that Saudi national interests have become more closely tied to the United States. This is true in terms of Saudi dependence on U.S. military and economic assistance, and Saudi interest in protecting their large U.S. investments. One could say that the interdependence is mutually beneficial. The Saudis have, in effect, agreed to protect the U.S. supply of oil in return for long-term military, industrial, and financial guarantees. In short, these mutual dependencies effectively explain the consistent declaration that Saudi Arabia will never again resort to the oil weapon.

SAUDI ARABIA AND ARAB POLITICS

Since closer U.S.-Saudi relations apparently rule out the future use of the OAPEC oil weapon, it may be said that the Arab world has entered a new era. Since the oil price increases of the early 1970s brought new wealth and political power to the conservative bloc, the Arab world has been edging toward a political stability unknown in the region since the beginning of this century. Consequently, this era has witnessed the general retreat of the radical cause in Arab affairs.

In Syria, the leadership of Assad has led to a purge of radical elements and unprecedented "forced" stability. In Oman, the Dhofari Revolution has become a thing of the past. Even within the Palestine Liberation Organization (PLO), the influence of the formerly uncontrollable radical elements has become insignificant. In Lebanon, the radical Muslim left has been effectively erased.

Simultaneously, Arab monarchies and military dictatorships, even the radical ones, have consolidated their power. In a region characterized by political instability which produced at least one coup or change of leadership annually for nearly 18 years prior to 1970, there has been only one successful coup in the last 15 years. Ironically, the single coup in Yemen brought that small state even closer politically to Saudi Arabia.

The retreat of the radical cause, coupled with Saudi power and the accompanying conservative resurgence, makes it difficult to say that oil will ever be used to achieve Arab foreign policy goals in the future, especially given the present situation in the Arab Middle East. The future use of the OAPEC oil weapon depends on the will of Saudi Arabia's leadership. Perhaps the only way the Saudis could lose their dominant position would be if peace negotiations stagnate. In this instance, the door would be open for a resurgence of radical popularity, thus placing tremendous pressure on the Saudis. It is doubtful, however, that the radicals can revive themselves without a total failure of Saudi policy or a devastating metamorphosis in the world energy situation. To hope for either is self-defeating for the entire Arab world. It seems quite remote that the balance of power in Arab affairs will shift away from Saudi Arabia in the near future. For this reason, OAPEC oil may not be used politically as long as the Saudi position remains unchallenged (see Table 3.3).

TABLE 3.3. Middle East Nations Compared: People, Area, and Revenue from Oil

Country	Area, Square Miles[a]	Population, Millions,[b] 1982	Government Revenue from Oil in Millions of U.S. Dollars[c]				Government Oil Revenue $ Per Capita, 1982
			1950	1965	1979	1982	
Bahrain	231	.37	2	17	780	1,300	3,514
Iran	636,367	40.24	45	522	19,090	17,250	429
Iraq	167,568	14.00	30	375	21,290	10,100	721
Kuwait*	7,780	1.56	12	671	16,780	8,600	5,513
Oman	82,000	0.95	–	–	2,000	4,100	4,316
Qatar	4,000	0.26	1	69	3,590	4,000	15,385
Saudi Arabia*	873,972	9.68	110	655	58,650	75,800	7,831
United Arab Emirates	33,378	0.79	–	33	12,860	15,500	19,620
Total Gulf Area	1,805,296	67.85	200	2,342	135,040	136,650	2,014
Egypt	386,198	44.67	2	40	760	2,100	47
Total Middle East	2,191,494	112.52	202	2,382	135,800	138,750	1,233

*Includes one-half of Neutral Zone.

Sources: Exxon Background Series, Middle East Oil and Gas (New York, December 1984). Reprinted with permission of Exxon Corporation, 1251 Avenue of the Americas, New York, N.Y. 10020.

[a]The Times Atlas of the World.

[b]International Monetary Fund, International Financial Statistics.

[c]Data based on exports of crude, refined products, and natural gas liquids.

NOTES

1. Arthur N. Young, *Saudi Arabia: The Making of a Financial Giant* (New York: New York University Press, 1983), p. 125.

2. Muhammad Loufti, "Prospect for Development and Investment for Oil Producing Countries," in *The Middle East: Oil Politics and Development*, ed. John Anthony (Washington, D.C.: American Enterprise Institute, 1975), p. 70.

3. Hans J. Morgenthau, *Politics Among Nations* (New York: Alfred A. Knopf, 1948), p. 15.

4. Ibid.

5. Malcolm H. Kerr, *The Arab Cold War* (New York: Oxford University Press, 1971), p. 107.

6. Saudi Arabia and Libya provided an annual subsidy to Egypt as a result of decisions made at the Arab summit in Khartoum on September 2, 1967.

7. Both the Egyptian and Jordanian responses to Jarring's many questions were identical. For more details see U.S. Senate Committee on Foreign Relations, *Chronology and Background Relating to Middle East*, pp. 255–75.

8. Don Peretz, "Energy: Israel, Arabs and Iranians," in *The Energy Crisis and U.S. Foreign Policy*, ed. J. Szyliowicz and B. L. O'Neil (New York: Praeger, 1975), p. 106.

9. The Egyptian People's Assembly formally dissolved the Egyptian-Soviet treaty.

10. General Nagib was the original leader of the Egyptian officers' coup in 1952. The Amin brothers had been accused of being CIA agents during Nasser's rule.

11. Obviously, Saudi Arabia's active commitment was absolutely essential to the embargo.

12. Nadav Safran, "The War and the Future of the Arab-Israeli Conflict," *Foreign Affairs* 52, no. 2 (January 1974): 219.

13. Sheikh Ahmad Zaki Yamani once said that the 1967 oil embargo "hurt the Arabs themselves more than anyone else . . . the only ones to gain anything from it were the non-Arab producers." Zuhair Makdashi, *The Community of Oil Exporting States* (Ithaca, N.Y.: Cornell University Press, 1972), p. 82.

14. Safran, "War and the Future," p. 210.

15. Ibid.

16. Edward F. Sheehan, "King Faisal Is Rich, Powerful, Determined," *New York Times*, January 26, 1975.

17. Ibid.

18. Joseph Kraft, "The Push Comes from the Middle East," *Washington Post*, December 9, 1976.

19. Henry Tanner, "Power Bloc in Arab World," *New York Times*, November 11, 1976.

20. Only Syria, Egypt, Lebanon, Kuwait, and the PLO were invited by Saudi Arabia to attend the Riyadh meeting.

21. Tanner, "Power Bloc."

22. Saudi Arabia controls 44 percent of OAPEC's oil.

23. John Duke Anthony's testimony before the U.S. Senate Committee on Foreign Relations, *Middle East Peace Prospects*, Hearings before the Subcommittee on Near East and South Asian Affairs, 94th Congress, second session, 1976, pp. 208–20.

24. Concerning the bargaining power of OAPEC vis-à-vis each other, see Douglas R. Bohi and Milton Russel, *U.S. Energy Policy* (Baltimore: Johns Hopkins University Press, 1975), pp. 60–70.

25. The conservative bloc's surplus capacity exceeds the radicals' normal total productive capacity. Moreover, a radical initiative is highly unlikely due to these states' short-term dependence on oil revenues.

26. Sheehan, "King Faisal Is Rich."

27. Jim Hoagland, "Saudi Leader Admits Israel's Right to Exist," *Washington Post*, May 25, 1975.

28. Emile A. Nakhleh, *The United States and Saudi Arabia* (Washington, D.C.: American Enterprise Institute for Public Policy Research, 1975), pp. 51–52.

29. Ronald Koven, "U.S. Saudis Sign Set of Agreements," *Washington Post*, June 9, 1974.

30. Don Oberdorfer, "U.S.-Saudi Bond Being Tested by Oil Cost Decision," *Washington Post*, December 12, 1976.

31. William A. Stoltzful's testimony before the U.S. Senate Committee on Foreign Relations, *Middle East Peace Prospects*, op. cit.

32. Ibid.

4

Islam Embattled:
The Iran–Iraq War

The Iran-Iraq war has many deep-rooted causes. At least seven factors precipitated the war: territorial disputes, ethnic dissent, linguistic chauvinism, religious differences, ideological contests, economic struggle, and personal vendetta. When war broke out in September 1980, adding new dimensions to the already turbulent Middle East, many analysts predicted a short campaign ending in mutual exhaustion, economic difficulty, and lack of ammunition. This forecast did not prove correct. The end of the war, now six years old, is nowhere in sight.

ANTECEDENTS OF THE CONFLICT

The Iran-Iraq struggle has manifested itself in various conflicts between the two countries. There has always been some conflict between the two states, such as the Shatt al-Arab and other border disputes, the issue of Iranian nationals in Iraq, Kurdish problems, and the arms race.

There have been border disagreements in the area throughout recent history. Although many have been resolved, several still remain. For the most part these disputes pose no major threat to either party or to oil resources in the area.

The case of Iran and Iraq is different from most other Middle East problems, for it is not a question of the small nations improving their

This chapter revises and updates an earlier article entitled "Holier Than Thou: The Iran-Iraq War." Reprinted by permission of *Middle East Review*, Vol. XVII, No. 1, Fall 1984. Copyright 1984 by American Academic Association for Peace in the Middle East.

status or economic gain. These issues are locally important, but region-ally and globally marginal. Instead, the territory in dispute may be valuable, and the issues go to the heart of leadership in the Middle East and in the gulf.

In 1975, Iraqi President Saddam Hussein and the shah of Iran signed the Algiers Agreement, establishing the navigable channel on the Shatt al-Arab River as the official border between the two coun-tries.[1] Iraq had long considered the entire river under its jurisdiction. The agreement granting Iran rights on the estuary was reached only after considerable pressure by the shah. President Hussein agreed to settle the Shatt border dispute in return for the shah's pledge to stop aiding Iraqi Kurds in their struggle against the Iraqi government. With the overthrow of the shah and the rise of Ayatollah Khomeini to power, the historical enmity between Iranians and Iraqis flared into war when Iraq moved its forces across the Shatt al-Arab border.

After the shah's deposition, the Khomeini regime immediately began a campaign against Iraq consisting of border confrontations and challenges, political provocations, and a much more dangerous campaign to destabilize Iraq through the support of Iraqi dissident groups.[2] The provisions of the Algiers Agreement regarding territory that should have been ceded to Iraq were not carried out. Iran con-tinued to treat the Shatt as sovereign Iranian territory.

Yet another problem between the two nations was the presence of many Iranians who had over a few centuries settled in and around the Shiite holy cities of Najaf and Karbala, in Iraq. These Iranians have not integrated in language, residence, custom, or nationality with the Iraqi population. They are viewed by many Iraqis as a potential danger to their society. Although the Iranians were a prob-lem during the days of the shah, the Khomeini government has further fomented their religious and ethnic sentiment. While there are few Iraqis in Iran, a significant number of people in Khuzistan are Arab. This in itself has been a bone of contention.

Over the centuries, Iran and Iraq have used Kurdish separatism against each other. Neither seems to like the Kurds, but each sees them as a people whose aspirations can be exploited to further their interests. Both nations have exploited the Kurds for a long time.[3]

The rivalry between Iran and Iraq has often translated into armed conflict. Since 1950 Iran and Iraq have armed themselves heavily. Since 1960 each has used the other's arms buildup to justify its own acquisitions. Iran, aligning itself for most of the period with the

United States, received arms superior to those provided Iraq by the Soviet Union. The arms race between the two nations has served to heighten tensions.

To further understand the conflict, one first has to look at the spread of Islam and Arab culture into Iran and other areas of the world, which began at the time of the Umayyad dynasty (A.D. 661–750). This purely Arab dynasty succumbed to the Abbasids, a Persian dynasty. In the meantime, the Prophet Muhammad's son-in-law, Ali, who was the fourth caliph of Islam, met with a violent death. Ali's son Husain, who was to succeed him, was also killed (A.D. 680). Out of the battle for succession emerged a sect in Islam known as the Shiites. Martyrdom is at the core of the Shiite creed. Shiites seek to avenge the martyrdom of Husain, whom Sunnis trampled to death at Karbala.

The Sunnis emphasize consensus in making laws. Shiites have developed the concept of the awaited *mahdi*, or messianic leader who will be empowered to reinterpret the Koran.

Still, the rift between the two major Islamic sects was incomplete until the emergence of the Ottoman Empire in the fourteenth century and the rise of the Safawid dynasty in Iran. The Safawids backed the Shiites, while the Ottomans took over responsibility for the predominant Sunnis. It is this Shiite-Sunni schism that is at the heart of the religious and sectarian dispute between Iraq and Iran.[4] Neither of the warring nations would like to admit it, because Islam forbids Muslims killing Muslims, yet both countries are creating a zeal for a holy war in order to inspire and galvanize the masses in support of their war policy.

Second, the Ottoman-Iranian rivalry created instability and tension in the Middle East, resulting in the political autonomy of many tribes on the frontiers of the two countries. The ferocious struggle continued for control of tribes and ethnic groups in Iranian Khuzistan, which the Arabs call Arabistan.

This local dispute between Arabs and Persians became international once the outside powers took the offensive in the Persian Gulf. In 1546, the Ottomans attacked Basra, an Iraqi port, in order to defeat the Portuguese and break their alliance with the Persians. Despite Ottoman expansion, Portugal maintained its control over Arabistan. When Spain annexed Portugal in 1640, England and Holland made their entry into the gulf. Although France negotiated a treaty of friendship with Persia, the former could not establish a foothold in the area. As the Persian empire weakened, enmity between Britain and Russia developed in the gulf region. The basic principle of Anglo-Russian policy in the Middle East was to maintain the area as a buffer to stop further rival imperialistic expansion. "This common interest," Tareq Ismail notes, "brought them into cooperation and collusion in settling Ottoman-Persian disputes, even while they competed and conspired against each other."[5]

Third, underlying the rivalry between Iran and Iraq are linguistic and cultural differences—a sense of superiority on one side and an inferiority complex on the other. Iran is inhabited by Farsi-speaking Indo-European people. Iraq is primarily composed of Arabic-speaking Semitic people. Muslims revere the Arabic language—the language of the Koran. No such reverence is held for Farsi. Iraq has repeatedly claimed Arabistan, an Arabic-speaking Iranian province, as its own territory. Iran has accused Iraq of cultural chauvinism. The Iraqi feeling of superiority, which emanates from the seventh-century Arab conquest of Iran, still lingers. Because Islam originated in Arabia and the language of the Koran is Arabic, many Arabs, including Iraqis, consider Iranians and other non-Arab Muslims as inferior, second-class Muslims. The Persian-Iranians, who have an ancient civilization and culture, resent the Arab-Iraqi attitude.

As a fourth factor, Islam, instead of uniting the two countries, has divided them. As stated earlier, Iran's population is predominantly Shiite. The Iraqi population is split between Sunni and Shiite Muslims, but the Shiites are more numerous. Khomeini's followers regard him as a *mahdi*. Therefore, they say, he was entitled to call Iraq's President Hussein anti-Islamic and declare a holy war against Iraq. The ruling elites in Iraq, including Hussein, are mostly Sunni, and therefore strive to orient Iraq toward Sunni Islam. The Shiites periodically demonstrate their dissatisfaction with the Hussein government. This religious division has created a holier-than-thou attitude between the ruler and the ruled in Iraq, and between the elites and the masses of the two countries. The superiority/inferiority complex which dominates the sociopolitical scene is described by Stephen R. Grummon as follows:

Moreover, both Arabs and Persians remember that it was the Arabs who conquered Iran and gave Islam to the Persians, but that it was Persian civilization that took raw, desert Islam and refined and tempered it.[6]

A fifth factor is that Ayatollah Khomeini, long in exile in Iraq, was expelled by the Iraqi government in 1978 at the behest of the shah. From his second exile in France, Khomeini began to urge his Shiite followers in Iraq to overthrow Hussein, who was preaching the secular ideology of Baathism.[7] On returning to Iran in 1979, Khomeini called upon the Iraqi Shiites to join him in a revolution against the anti-Islamic policies of Hussein and his Baathist followers. Iraq viewed the success of the Iranian revolution as yet another attempt to destabilize the region and bring about the downfall of the Baathist government. As Hussein said in a speech:

> They [the Ottoman Empire] took turns on Iraq: Turkey goes, Iran comes. All this is done in the name of Islam. Enough; no more Turkey, no more Iran. . . . We will not accept anybody coming everyday with a new path that aims at dividing Iraq and dividing the Arab nation.[8]

Iraq sees its war with Iran as Arab nationalism facing a tough conflict with Iranian imperialism. The Baath, as a secular political party, views Islam as the cultural and spiritual source of inspiration for Arab nationalism.

Iran's humiliation is the sixth factor. This was considered to be Iraq's call to greatness. The Iraqi leader envisions leading the Arab world in the suppression of Iran. Hussein dreams of resurrecting the glorious days of the Arab empire, and establishing Iraq as the region's greatest power.

Iran's chaotic state in the aftermath of the revolution was considered an appropriate time to eliminate the dictatorship of the clerics in that country. The shah's Iran was seen as a conservative power seeking to expand Iranian influence by preserving stability in the region. By contrast, the new revolutionary regime is seen as a revisionist power fomenting revolution and unrest. Khomeini's Iran has used foreign policy to challenge the domestic order of its neighbors, and has been far more persistent in the use of propaganda, subversion, and force than the shah's government ever was. In this respect, the xenophobic dictatorship of the clerics and Khomeini's rule has led to even greater repression than existed under the shah's rule.

THE IMPORTANCE OF GULF OIL

Recent events in the Middle East serve as a reminder both of the magnitude of U.S. interests involved in that fragile political framework and of the level and diversity of threats that surround them. It was not so long ago that interest and vulnerability alike were universally recognized. Today, apart from news reporters, senior government officials, and the businessmen whose activities make them unwilling actors in the drama, apathy is high. Why is the gulf so important?

What thrust the gulf into prominence was the awareness, born of the 1973–74 oil crisis, of Western oil dependence. It is not strictly true to say that this phenomenon was discovered in 1973. In Europe, dependence was recognized much earlier. Even in the United States, certain officials, aware of the growing dependence on Arab oil, tried in vain to awaken others to this development and its implications. What the oil crisis, and at least one later crisis, did was to demonstrate that dependence by giving us long gas lines, layoffs, inflation, and a sense of dislocation.

One of the frequent by-products of crises is a focus on the apparent rather than the real. So it was with the oil crisis of the 1970s. The image of gas lines, the anger and helplessness, and other less visible effects of the crisis led people to gauge U.S. defenses against future threats by watching the percentage of U.S. imports from the gulf, the level of stockpiles, the price of oil, the effects of conservation, and other statistics.

It is difficult to estimate gulf oil imports because gulf oil refined in Europe, Canada, Puerto Rico, and elsewhere is imported into the United States. Without gulf oil, though, there would be a serious shortfall, and the United States must compete for its supply on equal terms with other importers. In addition, the United States is committed to sharing in times of shortage with some of its allies which remain almost totally dependent upon gulf oil.

Reserve stocks rise and fall cyclically. It is difficult to keep abreast of them or their significance without a clear understanding of seasonal fluctuations in demand, government-versus-commercial stockpiles, agreements among downstream operators, and so forth. Stockpiles must be replenished, and while they provide some short-term insulation from oil crises, they are irrelevant to long-term problems.

Oil price is a function of demand and supply, retailing and cost considerations, and legislation and market forces. The decisions of OPEC on pricing are frequently undermined by a variety of factors. The weakening of OPEC in recent years reflects the diversity and power of market forces.

Conservation, reinforced by the extent of the 1981–83 world recession, has had a significant effect on Western oil demand. The result of the two forces has been to restrict growth in consumption rather than reduce consumption. The strong economic recovery can be expected to restore some growth in demand. Thus, while conservation has had a definite impact, it has not altered the dependence on gulf oil, which still makes the difference between an adequate and inadequate supply of energy.

Conservation measures are also irrelevant because they only affect the United States, as if its needs were somehow independent of the rest of the world. It has already been noted that existing agreements now require sharing of international oil under certain shortage conditions. Thus, the United States is committed to aiding Europe in the event of a shortage resulting from gulf problems. Since Japan and parts of Europe receive most of their oil from the gulf, the impact of a shortage would be considerably greater than the effect on U.S. imports, even when indirect (gulf oil sent to the United States from non-gulf countries) supplies are counted.

World demand has also been somewhat reduced by the worldwide recession. While all energy-importing countries are pleased at the relative reduction in gulf imports, no one considers a world recession a reasonable price to pay for it.

The economic devastation of the West resulting from a prolonged absence of gulf oil cannot be taken lightly. It is certain that the United States would feel at least a part of this disaster. Even the beginnings of a crisis have caused sharp increases in insurance rates and are now causing an increase in the price of oil on the spot market, both of which will mean higher prices for consumers of oil products. It has taken a decade to overcome the effects of the last oil crisis, which involved only a minor disruption of oil supplies. A long-term embargo of gulf oil could prove far more devastating.

None of this takes into account the vital role of gulf oil in the viability of NATO. The destruction of European economies would itself seriously compromise NATO, but it is not only European economies that run on gulf oil; it is also the military of the United States overseas and European forces.

The war between Iran and Iraq has caused concern for the future of gulf resources. This is a more specific aspect of the competition for gulf dominance. Iran, with its larger population and greater economic requirements, wants to establish itself as a dominant economic power in the gulf. This is important to Iran because its oil reserves, though large, are not great enough to provide adequate, stable, and long-term revenue for its people. Saudi policies relative to oil pricing, now supported by Iraq, run contrary to Iranian interests. Iran seeks to maximize the oil price. If this action increases efforts to find oil substitutions, Iran will not be affected significantly since the results of these efforts will not pay until after Iran's reserves are depleted. To create this influence, Iran must be able to force other suppliers to reduce production and increase the oil prices.

Iran's leaders hope to dominate all of the smaller gulf nations and quell Iraq's economic influence in the gulf.

The real target of Iran's economic drive is Saudi Arabia. Ultimately, all cutbacks depend on Saudi Arabia, and Saudi policies and objectives conflict with Iran's because of the huge Saudi reserves of oil and money. Only by destabilizing Saudi Arabia and its subjugation of Iranian hegemony may Iran realize its economic objectives in the region.

GATHERING STORM

We have given seven explanations for the hostility between Iran and Iraq. Iraq's dissatisfaction with the Algiers Agreement of June 13, 1975, its involvement in Iran's ethnic uprising, the buildup of political tension and frequent border skirmishes between the two nations, and Khomeini's call for Hussein's overthrow all indicated increasing tension in the region. In November 1979, the Iraqi government asked Iran to abrogate the Algiers Agreement and return the border areas, including the Shatt al-Arab river basin, to Iraq. It also demanded autonomy for Iran's minorities—the Kurds and the Baluchis.

Faced with growing pressure from Iraq, the new Iranian government sought to restore ties with the United States. However, Iranian Prime Minister Mehdi Bazargan's meeting with U.S. officials aroused suspicion among radicals in Iran that the United States was conspiring to mount another coup similar to the one that brought the shah back to power in 1953. They struck back by occupying the U.S. Embassy in Tehran and, with the support of the Iranian government, taking U.S. diplomats hostage. Iran's confused and chaotic government, playing into the hands of the radicals, mobilized a crusade against "foreign enemies" in order to divert attention from domestic troubles.[9]

The hostage crisis, and the ensuing U.S. confrontation, temporarily united those political forces in Iran that advocated exporting the

Iranian revolution to the neighboring states—including conservative Saudi Arabia. Thus the Mecca mosque takeover of November 1979 was inspired by Khomeini, who encouraged radical elements within the gulf states to believe that the time was ripe for revolution. In pursuit of this policy, in January 1980, Iran organized a conference of revolutionary Islamic organizations and initiated a campaign to stimulate revolutionary zeal in the region.

The hostage crisis isolated Iran from its traditional friends—and caused many nations to fear further revolutions in the area. Although Iran's new president, Bani Sadr, sought to resolve the hostage crisis in order to remove U.S. misgivings and reduce Iran's isolation, militant Iranians prevailed upon him not to do so until the United States had been thoroughly punished. Inevitably, Iran's provocative policy not only created internal chaos, ethnic unrest, and army plots to overthrow the government, but also isolated the nation and weakened its military capabilities.

With time seemingly in its favor, Iraq chose to strike, calculating that since Iran was diplomatically isolated, politically fragmented, militarily weakened, and without access to U.S. arms and spare parts, it would immediately fall to Iraq. But Iraq had grossly miscalculated Iran's resourcefulness, and Hussein failed to achieve a quick victory.

Several other facts have clearly emerged from this war, now in its sixth year. The superpowers have shown great reluctance to intervene directly in the war. The United Nations, the Organization of Islamic Conference (OIC), and the nonaligned movement have all failed to bring about a cease-fire. The Gulf Cooperation Council (GCC) has also attempted to serve as a mediator, and the foreign ministers of Kuwait and the UAE traveled to Baghdad and Tehran seeking to initiate talks. They, too, were unsuccessful.

All such efforts have foundered because of Iran's insistence that any resolution identify Iraq as the aggressor and require that Iraq pay substantial reparations to Iran. In order to induce Iran to accept a cease-fire and negotiations, Iraq has intimated that it will step up attacks on Iranian oil facilities and tankers transporting Iranian oil through the gulf. In response, Iran has repeatedly claimed that if its oil exports are substantially reduced it will seek to prevent any other country from exporting oil via the gulf by closing the Strait of Hormuz.

THE OIL WAR

The Iran-Iraq war has renewed concern about Iran's long-standing but ambiguous threat to close the Strait of Hormuz. The 26-mile-wide vital waterway is used by tankers that carry about 20 percent, or approximately 8 million barrels, of the noncommunist world's daily oil supply. Iran has repeatedly warned that it will block the strait (which lies at the mouth of the gulf) if foreign powers get involved in the war, or Iraq tries to cut off Iran's oil exports. The possibility of Iran's closing the strait cannot be ruled out, and this could precipitate another global energy crisis. This may provide the opportunity for other OPEC members to raise oil prices yet again.

The six Arab nations making up the GCC sit on one-third of the world's oil reserves.[10] Recent attacks on oil tankers have not only increased the cost of insurance, but have also made its availability more difficult. Air attacks on commercial ships have prompted the group to seek better security through diplomacy, as yet with no results. If the crisis continues, there may be pressure on the United States to intervene militarily to prevent Iran from closing the strait. President Reagan declared at a press conference on February 22, 1984, that there was "no way that we [the United States] could allow that channel to be closed."

Even if the fighting were to end, it would take about a year to repair damage to oil producing, pumping, and shipping installations in both countries. The restoration of refineries will take even longer. The economic effects are sure to be felt. The soaring oil price would renew inflation and increase the prices of other commodities, goods, and services. While this would hit both industrialized nations and developing countries, the latter would suffer more, as they have in the past.

It is a sad fact that those burdened with heavy debt would be most vulnerable. Although the Third World debt results partially from imprudent borrowing practices and flawed internal policies, the more fundamental explanation of these Third World debts lies in the energy crises of the 1970s and the unexpectedly rapid disinflationary process triggered by the antiinflationary measures adopted by industrialized nations.[11] A third energy crisis would be disastrous for most non-oil-producing Third World countries, precipitating widespread discord and leading to governmental changes and authoritarianism. Should another oil crisis occur, Third World leaders are sure to adopt aggressive foreign policies as a means of laying the blame for their internal problems on foreign scapegoats. Rising tension will be the certain outcome of such policies.

PHASES OF THE WAR

The Iran-Iraq war can be divided into several phases for a better understanding of how the parties arrived at the current situation. There is some debate between the parties as to what precipitated full-scale hostilities. One thing not in doubt is that the full-scale hostilities began in late September 1980 when Iraq invaded Iranian territory.

Iraqi forces concentrated on the southern Khuzistan front around Khorramshahr and Abadan. This was a strange choice because the most important targets in Khuzistan are north of Khorramshahr. In any case, Iraqi forces moved quickly into Iran. Iranian resistance was relatively weak. However, Iraq failed to move quickly enough and was unable to fight effectively at Khorramshahr. Iran attempted to stall the Iraqi advance until the rainy season, thus limiting Iraqi triumphs to symbolic victories.

In 1981, following the Iraqi stall at Khorramshahr, a number of Iranian counterattacks took place. The very fact that Iran was now attacking indicated that the initiative had passed to Iran, that Iraq was stuck in a war of attrition it could not win. This was the precursor to the next phase of the war. Iran began to mount effective offensives.

After the defeats of 1981, Iran's government and military underwent some changes. By 1981, the Iranian army had established responsibility for the planning and conduct of operations, had built a considerable base of experience, and was given much more autonomy. This change resulted in a series of Iranian victories over Iraq. By the fall of 1982, Iran had carried the war to Iraqi territory.

When Iraq began defending its own territory, its army's effectiveness improved. This was probably a result of Iran's inability to push beyond its frontier, or it may have been because Iraq was now defending the motherland rather than invading a foreign country. This helped Iraq to try to bring about a regional crisis of such severity that outside powers might intervene to impose a peace settlement. The tactic was to acquire the Exocet missile and the Super Etendard aircraft. These could be used to threaten oil shipping to and from Iran.

The friendly conservative Arab states have furnished Iraq with $35 billion to buy military weapons and supplies. Their aim is to arrest the revolutionary fervor and military influence that Iran had hoped to spread over the region's weak oil-producing nations. When it became evident that the aid had been inadequate, the Hussein government secured from France a fleet of Etendard bombers (armed with Exocet missiles) to blast away at tankers docking at Iranian oil installations on Kharg Island. Iraq's own oil exports had been virtually cut off by the massive Iranian assault on its ports. Iraq felt impelled to retaliate against Iran's oil trade. Iran has so far managed to finance the war by selling oil overseas.

Iran threatened to escalate the war by blocking the Strait of Hormuz. Giant tankers exit every hour through the strait, carrying at least 7.5 million barrels of oil every day. The United States sternly warned both sides not to disrupt international shipping, and it threatened Iran with serious reprisals if any attempts were made—by sea or air attacks, or the sowing of mines—to close the gulf tanker lanes. Carriers from the U.S. Sixth Fleet have been deployed in the Indian Ocean to police the area, and extra U.S. weapons were rushed to Saudi Arabia and Kuwait.[12]

STALEMATE IN THE WAR

The continuing stalemate can be attributed to four factors. First, though they possess large armies and air forces, neither Iraq nor Iran has been able to defeat its adversary or occupy major areas of the other's territory.

Second, the oil glut has had a dampening effect. OPEC finds itself left with unsold oil, largely due to the world trade recession and less consumer demand. There is more than enough oil to replace the supplies coming from the gulf immediately, and the oil-producing countries worry anxiously about their future revenues.

Third, the strife between Iran and Iraq has inflicted considerable suffering and casualties, and economic dislocation will become severe if the war continues. The governments of both countries have become more brutal with their own citizens and more threatening to the stability of their neighbors.[13] Massive arms transfers by the Soviet Union to Syria and Iraq, especially after the defeat of the PLO in Lebanon, have produced deeper divisions in the Arab League.

Jordan has joined Iraq and is more determined to see a new Iraqi oil pipeline connect with Aqaba than to deal with Israel over West Bank problems. Saudi Arabia, once the area's leading power, has had to scale back its ambitions in order to attend to its own security. The six GCC states have clamped down on subversion financed by Iranian Shiites, but they fear disaster if Iran continues to harass their oil tankers or wage heightened trench warfare against Iraq.[14]

Finally, the gulf states have actively discouraged the superpowers from deploying military forces in the region. On one side they refused to allow the United States to base the CENTCOM (United States Central Command) which was to be brought in as a permanent garrison. On the other side, the Soviets were warned against antagonizing Iraq or Saudi Arabia by throwing their weight behind Iran and Syria.

What is needed is a means of limiting war damages and speeding political conciliation. Every attempt has failed so far. The good offices offered by the United Nations and the Arab League have been declined by the warring countries. Diplomatic intervention by outside powers—including France, Britain, the United States, and various Third World nations—has been turned aside. It seems that the war may drag on until the combatants exhaust each other. In this war, no matter how many friends of interested parties try to intervene, it will end in disaster.[15]

SUPERPOWER STRATEGY

The war presents a puzzle, appearing to be self-contradictory, chaotic, and ambiguous to the superpowers. The United States does not have diplomatic relations with Iran, has professed neutrality since the beginning of the war, and does not supply arms either directly or indirectly to Iran or Iraq. While the two sides slug it out amidst minefields and trenches, however, the United States is committed to freedom of navigation in the gulf, a matter of vital importance to the world.

An occasional ally of Iraq, the Soviet Union has also maintained neutrality while facilitating the supply of arms to both sides. If it is Soviet strategy to seize warm water ports in the Indian Ocean, thus attempting to control the Strait of Hormuz or the oil fields, the Soviets are playing a risky game. As Joshua Epstein, in an article titled "Soviet Vulnerabilities in Iran and the RDF Deterrent," maintains, "the Soviets themselves seem to have appreciated these difficulties in a battle assessment of the region as long ago as 1941."[16]

Despite its proclaimed neutrality, U.S. policy toward the war and the region quickly emerged. This consisted of attempts at containment, calls for cessation of hostilities, continued access to oil fields, and efforts to avoid a U.S.-Soviet confrontation. For the United States, the war could not have come at a more opportune moment. It was in the midst of the Iranian hostage crisis. Although the war interrupted secret negotiations between the United States and Iran, U.S. Deputy Secretary of State Warren Christopher met in Bonn with Sadegh Tabatabai, who had connections with Khomeini. Tabatabai apparently gave a favorable report to Khomeini on the prospects of a negotiated settlement of the hostage crisis. Taking advantage of Iran's difficult situation, the Americans were able to negotiate the hostages' release from a position of strength: Iranian demands of $24 billion to cover their frozen assets and property taken by the shah and his family was cut to $9.5 billion. Iran paid $3.67 billion in outstanding loans with Western banks.[17]

The war came as welcome news to the people of the United States, who had helplessly watched the terrible hostage drama. Washington, however, remained uncommitted, both because of the hostage crisis and its desire to avoid Soviet intervention. It agreed, however, to supply AWACS and Stinger missiles to Saudi Arabia, along with the military personnel to operate them. This was considered by Iran as favoring the Arab side of the war, although by procuring these sophisticated weapons, Saudi Arabia was attempting to prevent Iran from attacking its oil fields. What had prompted the United States government to deliver the AWACS and the Stingers to Saudi Arabia was the imminent danger of an expanding war in the gulf. Iraq reportedly had moved some of its commandos and helicopters to Saudi Arabia, the UAE, Oman, Kuwait, Jordan, and North Yemen—all friendly Arab countries—and were preparing to attack the Iranian port of Bandar Khomeini. U.S. pressure on Saudi Arabia and the other Arab states forced Iraq to withdraw its forces.[18]

Linked to this U.S. policy was a desire to secure the Western and Japanese oil supply, and protect the supply route from the Soviets. With this objective, the United States has reinforced its naval presence in the region, aiming at a possible—but unlikely—blockade.

The United States has acquired military facilities in Oman, Somalia, Kenya, and possibly Egypt and Saudi Arabia, for deployment of CENTCOM, and has conducted training and joint exercises. Few analysts believe that CENTCOM can stop a major military move into the gulf region. Despite geographic and logistic problems, the Soviets could probably sweep aside any resistance should they decide to move into the region. A major deterrent to such an action is the economic cost involved in keeping the military operation there in the face of local resistance and guerrilla action. This move, in conjunction with the Soviet occupation of Afghanistan, would be a great financial burden. The political costs would also be heavy, for this would further alienate the Muslim community.

For its part, the Soviet position on the Iran-Iraq war was put forth by President Leonid Brezhnev in a declaration made in India. The principles of this declaration are:

1. Not to set up military bases in the Persian Gulf and on contiguous islands, and not to deploy nuclear or any other weapons of mass destruction there;
2. Not to use or threaten to use force against the Persian Gulf countries, and not to interfere in their internal affairs;
3. To respect the status of nonalignment chosen by the Persian Gulf states, and not to draw them into military groupings of which nuclear powers are members;
4. To respect the inalienable right of the region's states to their natural resources; and
5. Not to create any impediments or threats to normal trade exchange and the use of maritime communications concerning the states of this region with other countries.[19]

Like the United States, the Soviet Union can hardly be considered neutral, especially in view of its readiness to offer arms to Iran and its supply of weapons to Iraq (although it is likely that the Soviet offer of arms to Iran reveals its fear that Tehran might lean toward the West under the pressure of its need for weapons). The proximity of the Soviet Union to the Middle East leads to important security considerations. According to Stephen R. Grummon:

A ring of states closely tied to the United States to the south of the Soviet Union could force Moscow to divert resources, particularly military resources, to its southern borders. On the other hand, weak and pliant states on the southern border mean that military resources can be channeled to other areas such as Western Europe and China.[20]

The Soviet Union has "taken an unexpected interest in Middle Eastern oil."[21] As far back as 1921 it attempted to establish its claim to northern Iranian oil. During the interwar period, it joined the West in obtaining oil concessions in the Middle East. In 1947, Joseph Stalin extracted a promise of oil allotment from Iran as a condition for withdrawing Soviet troops from Azerbaijan, but this fell through mainly because of American and British interference.

Recently, the Soviet Union, unable to meet growing oil demands, has allowed Eastern Europe to import oil from the Middle East. Eastern European countries have thus made oil deals with both Arab countries and Iran.

Should the United States engage in major military operations, the Soviets might provide support to Moscow's clients and take other actions leading to a superpower confrontation. Such an event would spur opposition in Europe and the United States. Some European critics say the United States relies on military approaches to the problems of the Middle East because, as Gregory Treverton writes, ". . . those are easier in domestic politics than, for example, an attempt to move forward on the Palestinian issue."[22]

THE WAR'S IMPACT ON THE COMBATANTS

The war is now six years old. What are the combatants doing, and where are they going? Nobody knows. Both Iraq and Iran have given conflicting reports of the fighting, which sometimes follows months of relative inactivity. During these interludes, both sides prepare for further onslaughts. When they return to the battleground they launch massive air and missile attacks on towns and airfields across their borders.

The war has already been incredibly bloody. Some estimates put the number of dead at 500,000, many of them the result of human wave attacks by Iran's *Pasderan*—Khomeini's revolutionary guard. Iraq alone has lost 50,000, with another 50,000 wounded, and still another 50,000 taken prisoner.[23] It is possible that Iran, richer, bigger, and more populous, is gradually wearing Iraq down. Since the spring of 1983, neither side has been able to make progress on the battleground. Iraq has been in an increasingly difficult financial position. Its major option is to escalate the war in the hope that Iran will be forced to negotiate. Its oil route through the gulf has been blocked, cutting its revenues and forcing the Baghdad government to rely on support from Saudi Arabia and other Arab states. With its Super Etendard, received from France, Iraq wants to destroy Iran's Kharg Island oil complex, and has warned foreign tankers not to load oil there.

Iraq's fear of Iran seems to have exacerbated the Arab cold war and the divisions among Arab states. The six Arab gulf states, and Morocco, Jordan, and North Yemen have supported Iraq. A pro-Iranian camp, composed of the radical pro-Soviet regimes of Syria, Libya, Algeria, and South Yemen, has also emerged. This leaves the Arab world in disarray, lacking purpose, cohesion, and a sense of direction. The effects of the war on Islamic solidarity are shattering. The OIC, which had emerged as a force in international politics, failed to end the conflict.

It is ironic that pro-Soviet Iraq is being backed by pro-Western Arabs, while fundamentalist, theocratic Iran is strongly supported by the pro-Soviet Arab nations. Iraq once expected its ties with Arabism to be strong, but the war has shown this not to be the case. Even its relations with the Soviet Union seem to be fading. Iraq has reestablished formal diplomatic relations with the United States after a break of 17 years.

SOME IMPLICATIONS

The Iran-Iraq war has been underway for six years. In those years the gulf has not known peace, for not one of the countless cease-fire proposals has been accepted by both sides. Despite the importance of the war, the attention and concern of the outside world have been sporadic, punctuated by intermittent offensives and crises, broken by long periods in which eyes are focused on the world's other problems.

When the war began, world interest was focused on the gulf. Both sides boasted massive quantities of advanced military hardware. Both had very large armies. Neither side has had the logistical support to sustain intense combat. Since Iran broke the siege of Abadan, the war has been fought periodically. Iran has carried out assaults with negligible tactical results. The early pattern of Iranian withdrawal and advance, and the current stalemate have been characterized by suspense rather than offense. This is a large part of the reason for world inattention to the gulf war.

Another element has to do with oil economics. The war led to attacks on oil facilities, but neither producer's exports were as critical as those from other producers. Disruptions in shipping, except that of Iraqi oil, were relatively minor. Despite constant and growing threats to the oil trade, serious disruptions did not occur, thus reducing concern over the war.

Despite only occasional interest, the war has generated many important implications for the world. They can be classified as political, economic, social, and military.

Political. It is difficult to separate political implications from economic, social, or military implications. That the war has contributed to the polarization of the Arab world cannot be doubted. This is a reflection of the fact that the Arab states did not effectively solidify cooperation. They all saw common interests in defending themselves against radicals. The war increased Iraq's dependence on other Arab states, and improved its ties with Egypt, Saudi Arabia, and other moderate Arab states. Iraq's cooperation with moderate Arab states improved Iraqi relations with the United States and other Western nations. Diplomatic relations between Iraq and the United States have been reestablished. Another effect of the collaboration between Iraq and moderate Arab states is the demise of the coalition of Arab nations opposed to the Camp David agreement.[24]

The war seriously undermined Hussein's leadership. Despite Israeli aid to Iran, Iraq's position toward Israel came about in part as a result of the need for Western and moderate Arab support.

Saudi Arabia's role as leader of the moderate Arab gulf states has been reaffirmed, but rivalries among those states produced a clearer recognition of their vulnerabilities. This was particularly true after an Iranian-supported coup attempt in Bahrain.

The gulf war several times threatened the Strait of Hormuz. Each threat was met by an immediate and joint Western resolve to maintain freedom of navigation in the gulf.

Economic. The gulf war and its cost will have a major effect on the economics of the region. The billions of dollars used to wage the war could have been used for economic development. Since 1980, economic development in Iran and Iraq has been virtually halted. In addition, many of the war subsidies from other Arab states have been diverted from development and assistance to Third World countries. The postwar economic impact on both nations will be significant. The gulf war shows that, even in purely economic terms, the cost of modern warfare is always more than expected.

Another economic result of the war concerns the infrastructure in Jordan and Iraq.[25] In order to facilitate Jordanian logistic support for Iraq, the roadways between the two countries have been improved. With Iraqi ports closed to shipping, Iraq-bound cargoes were unloaded at Aqaba, Jordan, and transported overland to Iraq.

Social. The gulf war is symbolic of the rise of Middle East religious disputes. While there are many forms of unrest, sectarian conflicts seem to be creating even more instability than ethnic problems. Although the Iran-Iraq conflict is both religious and ethnic, there can be no question that its regional problems are primarily sectarian—a dispute between Shiite and Sunni Muslims.

Military. Before the war, the gulf had remained outside the realm of large-scale military operations, although arms procurement, training, deployment of forces and vessels, and threats and counterthreats have sounded the alarm for many years.

Before the war erupted, the gulf states had been talking about security for several years. The war's violence, U.S. naval activities near the gulf during the hostage crisis, the challenges presented by Iranian revolutionaries, and the Soviet invasion of Afghanistan gave greater impetus to the gulf security problem, leading to the formation of the GCC.

On one hand, the chances that gulf security will be threatened appear considerable. On the other hand, the ability of the Iranian and Iraqi armies to project sustained military operations beyond their frontiers is doubtful.

It seems that "the most difficult dilemma for Western policymakers would arise if the military situation turned against Iraq, and Tehran were able to bring about the replacement of Saddam with a regime subservient to Iranian interests."[26]

CONCLUSIONS

The Iran-Iraq war ostensibly involves issues on which well-meaning leaders can compromise. These issues are based on religious differences older than the two nations. They are fighting with the passion of a crusade.

The war has exposed the insecurity prevailing in the Persian Gulf region. Although the war is not about to end, the prospect of an Iranian victory would give Iran an edge over other states in the region.

Since Hussein chose the time to strike Iran, he was expected to show results. What is at stake now is his survival, and Iraq's world image. This failure to accomplish declared objectives could unleash an internal uprising in Iraq. In the wake of criticism of Iraq's ill-conceived policy, Hussein cannot afford to prolong the war. Nor can he stop fighting. Unless Iran accepts a cease-fire on Iraq's terms, there are no face-saving measures he can take. On the other hand, the longer the war lasts, the weaker his position becomes.

In Iran, the war has had a profound effect on the revolution. It has helped dissipate Khomeini's opposition and consolidated the dictatorship of the clerics. Internally, a curious balance between Iran's ethnic and ideological groups has emerged. Externally, Iran came out of isolation. Against all odds—isolation, a lack of relations with major powers, a U.S. arms embargo, domestic unrest—Iran has courageously faced Iraq's challenge.

Even when the war ends, Iraq-Iran irredentism will continue to be an important regional flashpoint, unless a Shiite administration takes over in Iraq.

The Iran-Iraq war has shown, beyond any doubt, that weak and insecure Third World nations, despite possession of vital resources, may not succeed in involving the superpowers in their contests. Despite the strategic importance of the gulf, both superpowers reacted to the conflict with restraint, if not perfect neutrality. Neither intervened nor became trapped in a war between two countries.

The Iran-Iraq war has proved that an aggressor may lose international prestige if it does not achieve a quick victory. Although Iraq denies any territorial ambitions regarding Iran's oil province of Khuzistan, the Iraqi offensive has curbed production there. Iraq has been known to use chemical warfare.

The war has demonstrated that the oil-consuming nations can absorb a loss of oil from selected countries. Unless an effort is made by OAPEC, no oil crisis or consequent price escalation can take place. Contrary to predictions, the war has encouraged many OPEC and non-OPEC nations to overproduce oil, causing a drop in prices.[27]

The war has offered the United States and its allies in the International Energy Agency further reason to cooperate in building more strategic reserves. A healthy competition in exploration, production, and procurement of oil has developed in the world.

Given their basic objectives in the region, both superpowers will try to forge closer ties with their client states. Washington will spare no effort to obtain more military facilities in Saudi Arabia and other gulf states, and Moscow will attempt to develop a stronghold in the region.

The Palestinian cause has suffered a further setback because of the war. Neither Iraq nor Iran can provide assistance to the beleaguered Palestinians. The war has pushed the Arab-Israeli conflict off center stage.

Finally, the war is bound to initiate policy changes in Iran and Iraq. Whatever direction policy takes, because of the long-standing nature of each country's claims and counterclaims, the prospect for a permanent resolution of the conflict appears bleak.

NOTES

1. See the text of this and other related treaties in Tareq Y. Ismail, *Iraq and Iran: Roots of Conflict* (Syracuse: Syracuse University Press, 1982), pp. 60–68.

2. Claudia Wright, "Implications of the Iraq-Iran War," *Foreign Affairs* 59, no. 2 (Winter 1980/1981): 278–79.

3. Edmund Ghareeb, "The Forgotten War," *American-Arab Affairs*, no. 5 (Summer 1983): 63.

4. For further discussion on the Shiite-Sunnite dispute, see Lawrence Ziring, *The Middle East Political Dictionary* (Santa Barbara: ABC-Clio, 1984), pp. 74–77, 79–83.

5. Ismail, *Iraq and Iran*, p. 5.

6. Stephen R. Grummon, *The Iran-Iraq War: Islam Embattled* (New York: Praeger, 1982), p. 2.

7. The ideological foundation of the Baath party is based on unity, freedom, and socialism. When the party came to power in Iraq in 1968, the Baath ideology became the foundation of Iraq's foreign policy.

8. Quoted in Shirin Taher-Kheli and Shaheen Ayubi, *The Iran-Iraq War: New Weapons, Old Conflicts* (New York: Praeger, 1983), p. 121.

9. Adda B. Bozeman, "Iran: U.S. Foreign Policy and the Tradition of Persian Statecraft," *Orbis* 23, no. 2 (Summer 1979): 387–402.

10. The six countries are Saudi Arabia, Kuwait, the United Arab Emirates, Oman, Qatar, and Bahrain.

11. Sylvia Ostry, "The World Economy in 1983: Making Time," *Foreign Affairs* 62, no. 3 (Summer 1984): 536.

12. See the strategic and economic significance of the Middle East oil in J. E. Peterson, ed., *The Politics of the Middle East Oil* (Washington, D.C.: Middle East Institute, 1983), pp. 3–36, 103–43.

13. The cost of the war is noted in detail in the *Strategic Survey 1982–1983* (London: International Institute for Strategic Studies, 1983), pp. 63–84.

14. *Wall Street Journal*, May 2, 1984.

15. For details on arms sales and defense expenditures, see the yearbook published by the Swedish International Peace Research Institute, *World Armaments and Disarmaments* (New York: Taylor and Francis, 1983), pp. 267–390.

16. Joshua Epstein, "Soviet Vulnerabilities in Iran and the RDF Deterrent," *International Security* 6, no. 2 (Fall 1981): 126–58.

17. Sepehr Zabih, *Iran Since the Revolution* (Baltimore: Johns Hopkins University Press, 1982), p. 60.

18. *Newsweek*, October 13, 1980.

19. A. Alexeyev and A. Fialkovsky, "For a Peaceful Indian Ocean," *International Affairs* (February 1981): 87.

20. Grummon, *The Iran-Iraq War*, p. 63.

21. John A. Berry, "Oil and Soviet Policy in the Middle East," *Middle East Journal* 26, no. 2 (Spring 1972): 149.

22. Gregory F. Treverton, "Defense Beyond Europe," *Survival* 25, no. 5 (September/October 1983): 220.

23. *Plain Truth*, March 1984.

24. Grummon, *The Iran-Iraq War*, p. 63.

25. Nameer Ali Jawdat, "Reflections on the Gulf War," *American-Arab Affairs*, no. 5 (Summer 1983): 94.

26. Michael Sterner, "The Iran-Iraq War," *Foreign Affairs* 63, no. 1 (Fall 1984): 142.

27. *Durham Morning Herald*, May 28, 1984.

5

Muslim Insurgency in Afghanistan

During the Christmas Week of 1979, the world was shocked by grim events in Afghanistan. A massive invasion by Soviet armed forces overthrew the Afghan government and launched a campaign of violence and terror against the Afghan people. Since then, the doggedness of the insurgents and the muted satisfaction of the Russians, coupled with the fact that neither side has yet suffered a serious setback, lead us to believe that the real war in Afghanistan has barely begun.

Six years after its troops swept into Afghanistan, the Soviet Union is still fighting a ferocious and largely secret war to subdue a fiercely resistant people. The Soviet Union undertook this invasion to contain a rapidly growing insurgency against the oppressive radical Marxist government of Hafizullah Amin. The countrywide movement threatened to end the rule of a Marxist regime to which the Kremlin had become heavily committed and put in power a government uncontrollable by and probably hostile to the Soviet Union.

BACKGROUND

On December 24, 1979, Soviet airborne troops began to land at Kabul, the capital of Afghanistan. By December 27, the total of Soviet troops in Kabul had risen to 5,000. Under varying pretexts, they disarmed many Afghan troops stationed in the capital. That night, Soviet forces stormed Darulaman Palace, the residence of President Amin. Afghan soldiers loyal to Amin were overcome by the Soviets. At the same time, a transmitter in the Soviet Union, claiming to be Radio Kabul, broadcast a taped announcement by Babrak Karmal, one of the founders of the People's Democratic Party of Afghanistan (PDPA), that Amin had been overthrown by a coup of party members.

A few hours later, the real Radio Kabul, seized by Soviet troops in a coordinated attack, began broadcasting in the name of the new Afghan government. It proclaimed that Karmal had been named president of the Democratic Republic of Afghanistan (DRA). Shortly thereafter, it disclosed that President Amin had been tried by a party tribunal and executed. The same night Radio Kabul announced that the Soviet Union had accepted an urgent request from the Afghan government for military assistance.

Thousands of troops poured across the Amu Darya (Oxus River) from the southern Soviet Union into Afghanistan or arrived at airfields under control of Soviet forces previously dispatched as advisers. The influx mounted until early January 1980, when there were 40,000 Soviet soldiers in Afghanistan. By that summer there were 85,000. They occupied all major cities, enforcing Karmal rule on a land that has never long been held by foreign invaders.

The Marxist leadership that the Soviets altered in December 1979 had come to power in an April 1978 coup in which Prime Minister Mohammed Daud Khan was ousted and killed. Daud Khan was an early leader of the nonaligned movement, adept at balancing East and West. Daud Khan himself had deposed his cousin, King Zahir Shah, in 1973. Shortly after the Saur (April) revolution, the two factions making up the PDPA split, and Karmal's Parcham group was purged by the Khalq faction of Vice President Amin and President Noor Mohammed Taraki. In September 1979, following a Soviet-backed attempt to eliminate him in a shootout in the presidential palace, Amin declared himself president. The following month Kabul announced Taraki's death.

The communists in Kabul alienated the Afghan people by insensitive enforcement of social and economic reforms, including a land redistribution program which encountered immediate and intense opposition. Dissent was met with brutal repression.

Opposition to the communist government grew quickly and spontaneously throughout Afghanistan. Virtually all elements of the population were involved: Islamic fundamentalists, who had already organized in opposition to the king and prime minister; parliamentary moderates; royalists loyal to the king; army officers resentful of the growing role of Soviet military advisers; traditionalist and tribal elements angered by the government's efforts to enforce its programs in areas where the central government's writ had never run large. These ethnic and tribal groups, which are the rural or nomadic majority of the population, form the core of the resistance.

HISTORICAL BACKGROUND

Invasion is not new to Afghanistan. Historically, it has been a battleground for political and religious forces between Europe and Asia. This description is accurate in terms of the country's role as pathway for numerous invaders. Aryans invaded Afghanistan and drove out the inhabitants. Later invaders were Persians; Greeks; Kushans; and Arabs, who brought today's dominant religion—Islam. Britain and imperial Russia contended for control over Afghanistan in the nineteenth century. In 1839, Britain invaded Afghanistan to reduce Russia's influence in the region. After World War II, the Soviet Union, the United States, China, and more recently India, Pakistan, and Iran have assiduously contended for influence in Afghanistan.

Afghanistan is a landlocked country in southwest Asia. It is bordered by the Soviet Union, China, Iran, and Pakistan. This gives it a geographical and strategic importance far beyond that which its political and economic position merits.

By the end of the 1880s, southward tsarist expansion reached the Amu Darya, which thereafter was Afghanistan's boundary with Russia. In defense of their Indian empire, British troops occupied Kabul in 1879–80. With the boundary between Afghanistan and India demarcated in 1893 at the Durand Line, Afghanistan was removed from the zone of Anglo-Russian contention. Russia and Britain met in 1907, establishing Afghanistan as a buffer state. This situation continued even after the Russian Revolution of 1917. Afghanistan, as a neutral country, also escaped foreign military occupation during World War II.

INTERNAL DYNAMICS

Afghanistan is primitive by any definition, and the country is very remote, rural, and isolated. The Afghans are nearly all Muslims. They are divided into 21 ethnic groups, each of which has its own language and culture. Its political tradition of tribal allegiance—overcast by deep and mutual suspicions—makes progress difficult to achieve.

Afghanistan is an underdeveloped country. The annual per capita income in 1979 (year of Soviet occupation) was $100, compared with $7,100 in the United States.[1] Per-capita-income figures tend to minimize extremes of poverty existing among the Afghan tribes and rural population since much of the wealth remains in the hands of a small segment of the total population living mostly in urban areas. In addition, most of the people are politically underprivileged. This is partly due to long years of foreign domination, rule by absolute monarchs, and underdevelopment.

Article Two of the 1964 Afghan constitution declares: "Islam is the sacred religion of Afghanistan." The government had been significantly involved in supporting it. Islam is universally prevalent among the peasantry and tribesmen. Religious attitudes and practices permeate the lives of most people from birth to death. Their overall attitude toward life, marriage, divorce, and the universe are influenced by Islamic standards, precepts, values, and customs.[2] Traditional tribal codes in many areas of this mountainous country demand blood vengeance. *Mullahs* (Islamic theologians), along with tribal *maliks* (leaders) "hold power of life and death over their followers, and the government dare not interfere."[3]

To the mullahs and tribespeople, Muslim rulers rule because of divine sanction. This sanction can be withdrawn by the mullahs anytime the rulers go against the will of Allah. The mullahs refuse to accept abolition of the veil and they want isolation of women from men. The Afghan mullahs oppose secular education; they call land reform anti-Islamic, for they own large tracts of land in the name of *waqf* (religious endowment); and they oppose separation of mosque and state, for such a move diminishes their power and hold on the populace.

It is always difficult to blame one's own culture for failing to meet the challenges of the modern world.[4] It is much easier to point to some outside influence. Many people in Afghanistan assume, without question, that their miserable conditions and the inability to do much about them are the direct result of Western imperialism, colonialism, and feudalism. This resentment is never intense except among small groups of Afghan intelligentsia who exacerbate the country's domestic and international problems.

Afghanistan's internal situation in 1973 was dominated by serious difficulties arising from three successive seasons of drought. Famine was widespread and there were many deaths from starvation. Massive Soviet and U.S. aid programs, supplemented by efforts sponsored by Britain, China, France, and India, had done much to improve transportation, communication, irrigation, and other essential elements of economic substructure on which progress could be based.

The situation began to change after the overthrow of King Mohammad Zahir Shah in July 1973 by Daud, who announced his adherence to Afghanistan's traditional policy of nonalignment. Nevertheless, Daud was a friend of the Soviet Union. Members of Afghanistan's two pro-Moscow parties joined the government, and numerous Soviet advisors entered the country. Hundreds of Afghan army officers were sent to the Soviet Union for training. Many Soviet teachers taught in Afghan schools.[5]

Daud was a respected and prudent politician. After becoming president he began to redirect the nation's foreign policy. Relations with Pakistan and Iran were improved, displeasing the leftist groups that helped him return to power. Afghanistan and Iran agreed to construct a railway between the two countries and Iran offered $2 billion in economic aid to Afghanistan. The two countries also signed an agreement regarding the division of the Helmand River. Daud's attempt to establish cordial relations with Iran, then a member of the now-defunct Central Treaty Organization (CENTO), and lack of support for ethnic nationalism in Pakistan, another member of CENTO, seriously displeased the Soviet Union. At the same time Daud replaced many high leftist officials with officers loyal to him, and refused to share power with leftist groups. This move was implicitly directed against the Soviet Union.

Daud's failure to deal with the country's economic problems further contributed to his unpopularity. It is estimated that between 300,000 and 1 million unemployed Afghans left the country to find employment in Iran and other oil-rich Middle Eastern states.[6] Daud's downfall came in April 1978 when he lost the support of most organized political groups, including the communists. His own political party, the National Revolutionary party, failed to generate popular support for his government.

SOME EXTERNAL CONSIDERATIONS

The Soviet Union had pressured King Zahir Shah to adopt a pro-India policy during the India-Pakistan war of 1971. Daud, after his return to power, supported Moscow's Asian security plan and displayed hostility toward Iran and Pakistan. He supported Pakhtoon and Baluchi demands for independence. By 1976, however, Daud had agreed to respect Pakistan's territorial integrity. Relations with Iran improved as well. He visited Saudi Arabia, Pakistan, and other Muslim countries. Daud had realized that as a member of the royal family he could not be the ultimate choice of the Soviets. By coming closer to Muslim countries, particularly Iran and Pakistan, Daud made an effort to adopt an independent foreign policy and reassert the Afghan nonalignment. His statement that Afghanistan wanted "true nonalignment," while Cuba only claimed nonalignment, was likely to have annoyed Moscow. Daud was ousted in April 1978. Since then, there has been a succession of coups, the most recent in December 1979, bringing Babrak Karmal to power.

The Soviets claimed that they have been invited to help protect the Afghan government. President Amin and several members of his family were assassinated after the Soviets gained control of the capital city. Several days later Karmal was brought into power. Soviet and Afghan statements are contradictory, if not vague, about who made the fateful request for Soviet troops and the date on which it was made. Karmal's envoy to the United Nations, who defected after the coup, stated that Soviet aggression had been a unilateral decision, a miscalculation, and a mistake, which had seriously affected the traditional friendship of Afghanistan and the Soviet Union. He further added that the Afghan people did not know why their big neighbor, which always claimed friendship and peaceful coexistence, trampled Afghanistan. Even after six years, the Soviets are in great trouble in Afghanistan. "The Russians can no longer hope to establish a pro-Soviet government and army capable of surviving the departure of Soviet troops."[7]

The Soviet operation in Afghanistan has been a military humiliation and an unfortunate political adventure. The ferocity of Afghan resistance caught the Soviets by surprise. The Soviet Union admits that it made more mistakes than usual. "Unlike the Czechs in 1968, they [Afghans] have not been silent watchers at their country's funeral."[8] Despite the deployment of about 115,000 men, law and order have not been restored in Afghanistan. Guerrilla strikes and local rebellion have given way to an authentic war of liberation, the effects of which are being felt throughout the world. It has had a profound effect on international security.

The Soviet occupation of Afghanistan is regarded by some analysts as the first step toward warm-water ports in Iran or Pakistan. The region provides the Soviets with a corridor to the warm waters of the Arabian Sea and the Indian Ocean. It seems Moscow is determined to disregard all warning signals and press southward to the gulf. The Soviet Union can deploy troops and aircraft within easy striking distance of the gulf, only about 300 miles from Afghanistan. The newly constructed naval base at Chahbahar, Iran, only 50 miles from the Pakistan border, and Gwadar, Pakistan, lie close to the Afghan border.

It appears that the Soviets aspire to control the region south of their country in the direction of the Indian Ocean, the Arabian Sea, the gulf, and, through the Red Sea, the Mediterranean. This desire has never been overtly proclaimed by the Soviets, yet their geopolitical perception regarding access to the high seas and control of the oil region make their desire apparent. It is a fact that large oil reserves are found in only a few areas of the world, such as the Persian Gulf region. A number of observers believe that the rich oil-producing countries will interest many cooperating or competing parties in the future.[9] With Iraq's determination to be less dependent on the Soviet Union, the gulf port of Iraq, Umm al Qasr, may not be available to the Soviets. The Soviet Union can use Syria to prevent Iraq from escaping its orbit.

Moscow may be seeking to close gaps in its arc of influence stretching from the Horn of Africa to Central Asia. One end of this arc is anchored in Ethiopia, where Colonel Mengistu Haile Mariam's military government puts Moscow in a position to control the Red Sea and the Suez Canal. In late 1977, Somalia severed its ties with the Soviet Union and expelled all Soviets from the port of Berbera. The Soviet Union, therefore, increased its naval forces in the Indian Ocean during the 1977 Ethiopian campaign to recapture the Ogaden from the Somalian-supported Eritrean Liberation Movement. During the weeks immediately following Soviet intervention in Afghanistan, there were reports that both superpowers had begun strengthening their navies in the Indian Ocean. The Soviet Union also is boosting its influence in the Seychelles archipelago, 1,170 miles off the coast of Kenya. It is a strategic position on the route used by all freighters bound for the United States and Western Europe. The Soviets see the Seychelles as strategic and feel the island can be effectively used to balance the U.S. naval base on Diego Garcia. Some analysts believe Moscow is slowly moving to circle the Gulf of Aden-Persian Gulf-Arabian Sea region.[10]

PEOPLE'S DEMOCRATIC PARTY

The coup in which Daud was killed brought the PDPA to power, which is commonly known as Khalq (people's). The Khalq was established in 1965. Taraki, a founder and secretary general of the party, took power after the coup of April 1978. Many political and religious groups did not regard the Taraki government as legitimate. Taraki was known as a Marxist and nationalist. The Khalq party drew its inspiration from the Communist Party of India.[11] In the wider context, this affiliation involves both Soviet and Indian dimensions.

Although Taraki professed nonalignment, he was actually relying heavily on the Soviet Union for economic aid and advice. In December 1978, after Taraki visited Moscow, a 20-year treaty of friendship and cooperation was signed with the Soviet Union. This Afghan-Soviet treaty was similar to those signed by the Soviet Union and East European countries in the 1940s, and more recently with Vietnam, India, and Ethiopia. This development led many observers to believe that Afghanistan was becoming Soviet dominated. Many Soviet civilian and military advisors were brought into Afghanistan.

The Khalq government argued that Afghanistan's economic problems were caused by feudalism at home and capitalism in the West. It emphasized class struggle as a means of eradicating the evils of economic exploitation.

For a variety of reasons, including personality clashes and individual ambition, the PDPA was split into two factions. The seceding group became known as the Parchamis, because it published a newspaper called *Parcham* (flag). One of the founders of the Parchami group was Karmal, now president of Afghanistan.

During Daud's regime the Khalq-Parcham participated in parliamentary elections. They knew that with their socialistic program they were unlikely to win control of parliament in a conservative Muslim society. They began to use extraconstitutional means of taking power. The Khalq-Parcham infiltrated the armed forces and alerted comrades in the army to the proletariat class struggle.

The Khalq-Parcham began to organize large anti-Daud demonstrations. President Daud demanded that they disband their organizations and become members of his National Revolutionary party. This was opposed by the Khalq-Parcham coalition. After the leaders of these two leftist groups were arrested, their supporters in the army overthrew Daud and installed the Khalq-Parcham coalition to power.

When the Khalq-Parcham coalition came to power, a conflict arose over the policies of the new regime. The Khalq eliminated a number of top Parchamis from high positions in the government. An internal power struggle ensued. President Noor Mohammad Taraki was replaced on September 16, 1979, by another Khalqi, Amin. Amin had incurred Soviet wrath by failing to crush the Muslim insurgency that had developed in the country's tribal and rural areas. The tribal insurgents had taken to butchering Soviet advisors.

INSURGENCY

Since April 1978, when Afghan Marxists staged their coup, three presidents have died violently. President Karmal is continuing the Afghan practice of eliminating potential enemies. The Afghan Revolutionary Council formed by the government has called upon citizens to form militia brigades to fight the insurgents.[12]

The Afghan army, supported by Soviet troops, has been meeting stiff resistance from the insurgents and underground opposition. Resistance to the Soviet military presence and the Marxist government spreads from time to time to the major cities, including the capital. The insurgents, mullahs, and other rightists and nationalists organize joint protest strikes. The government imposes martial law and curfew to curb insurgent violence, plundering, and arson. The insurgents resort to assassinating family members of top government officials. The young militia of the ruling party arrests and executes insurgents in cities. The purge and counterpurge continue. These killings are not easily forgiven. Although the number of people killed is highly exaggerated, it is estimated that 100,000 civilians have been executed by both sides.[13]

The Afghan insurgents are motivated largely by vengeance. They are deeply divided along tribal, ethnic, and political lines. They are poorly armed. Their greatest weakness is the lack of a single charismatic leader. There is no Kemal Ataturk, Mao Tse-tung, or Ho Chi Minh to unite them.

Since the Christmas week of 1979, an estimated 80,000 Soviet troops and nearly 50,000 Afghan soldiers have helped the leftist government quell the Muslim insurgents who call themselves *Mujahidin* or Muslim holy warriors. A misunderstanding exists in the West that the Afghan insurgents are pro-West. This is not true. Their anti-Soviet stance springs from Islamic interests. Although not pro-Western, they are undoubtedly anticommunist.

Frustration is the lot of the Afghan insurgents. Because the tribes and ethnic minority and religious groups receive little or no political expression in the existing system, many turn to insurgency. For them, Islam provides an acceptable reason to launch a *jihad* or holy war against the Soviet-backed Marxist government. Beyond this allegiance to Islam and hostility toward Marxism, there is little agreement over a future political system for the country. The insurgents are seldom able to rally much support from other nationalists. Since the insurgents are essentially tribal groups, they can gain popular support of a particular clan. Their hostility toward the government is tribal and religious rather than national or secular.

Torn by tribal rivalries for centuries, Afghan insurgents are attempting to organize an alliance of six loose groups comprising some 60 tribes in order to qualify for financial aid from Muslim nations. The tribesmen have plans to select 100 members of a revolutionary council that could unify the fractious movement under a united front called the United Islamic Liberation Front.[14] No united front is likely to emerge, however. The tribes are fearful of subjecting themselves to a larger united group.

The insurgents in Afghanistan are mostly tribesmen, mullahs, and their followers. The insurgency is a result of tribal attitude toward religious and political frustration based on a feeling of helplessness with regard to: a growing conviction that the Soviet invasion of Afghanistan has turned into a full-scale occupation; a resentment of the government's control over the country and its determination to modernize a primitive people and society; the fear of loss of their religion, custom, and culture under an infidel and Marxist-controlled government; the anti-Islamic repression by the regime; the sectarian religious conflict; the proud record of repelling invaders; and the protection of their ancient tradition of rough-hewn democracy.

These conditions have resulted in apprehension and a search for alternatives. These alternatives are to apply selective violence and inflict psychological and political damage on the government. The violence is meant to loosen the hold of the country's Moscow-controlled government. Insurgent strategy is actively to resist the government's authority, disrupt communications, and disorganize life by damaging essential services. They hope to shatter the credibility of the rulers, reduce their ability to govern and collect revenues. Further, the insurgents desire to stir up unrest among religious groups with a view to sabotaging government decision making and unsettling the country's economy to hinder progress under the Marxist regime.

PHASES OF INSURGENCY

Insurgency is accompanied by frustration and unrest during its incipient phase, agitation and mobilization problems in the organizing phase, both violence and nonviolence during its operational phase, and liberating and legitimizing symptoms in the concluding phase (see Figure 5.1). The following is an examination of the symptoms and manifestations of each phase, revealing the evolution and process of insurgent strategy in Afghanistan.

The earliest symptoms of insurgency are individual frustrations. Tribal and religious leaders are the first to express discontent and provide direction for revolt. Northern Afghanistan is more rigidly anticommunist than the rest of the country because many of its people and religious leaders remember how the communists tried to stamp out Islam in the Soviet republics of central Asia in the 1930s, sending thousands of Uzbek and Turkoman Muslims into Afghanistan. The Soviet invasion of Afghanistan has only inflamed resentment that has been seething since the Marxists came to power.

Eastern Afghanistan is inhabited by fiercely nationalistic tribesmen. They have defeated and repelled every foreign invader. They have fought three wars against the British, the latest in 1919, which won Afghan independence in foreign affairs. The common goal of northern and eastern fundamentalist Muslim tribesmen is to expel

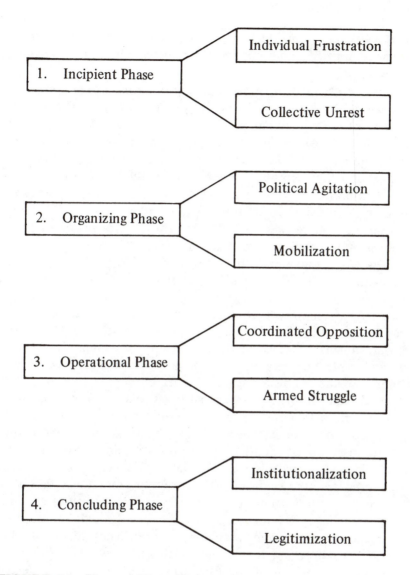

FIGURE 5.1. Phases of Insurgency

the Soviets and topple the Marxist government of Karmal. These two Afghan groups have the strongest feelings about religion, ideology, and nationalism.

The symptoms of frustration among the tribal groups and leaders are personal complaints and criticisms of those in power. The words of many of these unemployed and uneducated people, fraught with economic setbacks in recent years, smack of revolt. Another symptom is the tendency of the peasants and rural masses to find time for political activities. Normally preoccupied with making a living, they are not interested in politics. Dissatisfaction drives them to seek solutions which are rarely found through peaceful means.

Collective unrest is usually expressed through spontaneous demonstrations, strikes, and riots. In Afghanistan, such occurrences are beginning to take place on a larger scale. Most Afghans are content with the simple life and are alienated from elite-dominated urban politics. Tradition has a strong hold on them. Over the centuries Afghans have worked out methods of solving local problems. They have not experienced a harsh break from village life. Only those who are unable to adjust to the new reality of communist domination and control express their unrest through armed insurrection and mount sporadic ambushes against Soviet convoys. Sometimes they manage to launch hit-and-run raids into vulnerable Soviet-held cities.

They are seldom able to mount coordinated, large-scale military operations. The insurgents have been receiving material support from such fundamentalist political groups as Hizbe Islami, Ikhwani-Muslemin, and Jamiat-i-Islami. The Jamiat-i-Islami, as an insurgent group, operates from Pakistan. The surrounding mountains, adjoining the sanctuary in Pakistan, are a stronghold for the insurgents. Outsiders supplying arms reportedly include the United States, Iran, China, and Pakistan. Egypt is training Afghan insurgents in guerrilla warfare.[15] Saudi Arabia is helping financially.

An early symptom of the organizing phase is the beginning of agitation by professional revolutionaries among the frustrated and politically conscious. In Afghanistan, it was the Khalq-Parcham coalition which first worked among the proletariat and the oppressed. They were the leading revolutionaries against the unpopular governments of Mohammad Zahir Shah and Mohammad Daud Khan. The introduction of a communist regime in Afghanistan was inspired and supported by the Soviet Union. National leadership has been offered by both the Khalq and Parcham groups. These groups have had little success among the Afghan tribes, who are conservative, deeply religious, and anticommunist.

It is now the turn of the tribal maliks and mullahs to organize public support against the Marxist regime and Soviet occupation. Slogans urge national liberation. Other themes decry the inequality between the rural and urban populations, the unemployment problem, and the poor living conditions attributed to past colonialism and the present communist regime.

THEMES

In Afghanistan positive themes, like the negative, are effective only among dissident groups. Since most of the insurgents are tribesmen, it is extremely difficult for them to win the support of urban elites, promises notwithstanding. To overcome this handicap, the United Islamic Revolutionary Council, a pro-West group, was formed. But the council lacks the vigor of most revolutionary movements. Because of tribal and ethnic friction, the tribesmen are unable to organize effectively across racial lines. They have attempted to unify the heterogeneous elements under the banner of Islam. The ability of the insurgents to discredit the communist regime has been more successful than their ability to promote a positive program of building a new Afghanistan, as the Marxists have promised.

Being masters of the art of turning practical demands into holy causes, the tribal leaders transfer their people's opinion into simplified themes, giving the impression that they are reaching the heart of the matter. They devise symbols, both verbal and visual, to arouse prejudice of things unacceptable to the people. Name calling and derogatory labels are effectively used. Conversely, the use of flattery and glittering generalities prove equally effective. They devise instruments to exploit every possible means of promoting their plan.

Since the Soviet takeover, the insurgents have tried to form organizational bases from which to operate. The organizational symptoms of insurgency have become manifest since the occupation. The function of the mobilization is to step up agitation and propaganda, discrediting the puppet government and preparing the population psychologically to support the insurgents' operational phase.

With the mobilization of politically conscious groups, there emerges a substructure of propagandists, agitators, spies, and political leaders directing the dissatisfied elements of society. Opposition is expressed in well-timed, coordinated mass meetings, demonstrations, strikes, riots, and acts of civil disobedience.

The activities often begin with a mass meeting of tribesmen and peasants. Insurgents are not able to form groups of politically conscious intellectuals, students, laborers, and workers. Because in societies such as Afghanistan, "kinship replaces government and guarantees men and women born into a specific unit or functioning set of social, economic, and political rights and obligations."[16] The emphasis is on the individual rather than the group, although collective participation is not rare. Since the Soviet occupation, insurgents have been desperate to form organizational bases for their movements and liberation efforts. But, because of tribal rivalries, personal enmities among leaders, ethnic heterogeneity, and a lack of communication facilities, the insurgents have not been able to develop coordinated opposition to the communist regime. They almost never are able to implement a plan involving the whole country.

VENDETTA

The communists have been stirring up a vendetta against themselves since they seized power in 1978, killing or dividing many peoples and families. The communists have invited vengeance. "Blood for blood" is an ancient Afghan tribal law. The tribal insurgents have been using terror to intimidate and disrupt the administration. The insurgents are becoming ruthless. Almost indiscriminately, they terrorize citizens and kill government officials and soldiers.[17] Government properties are set on fire. Civil servants, office workers, shopkeepers, and the public join in civil disobedience and general strikes from time to time to paralyze the normal functioning of the administration. In response, the government resorts to arresting people for their suspected role in antigovernment and anti-Soviet riots. In the tribes there is very little governmental authority. Any indiscriminate killings or use of terror may alienate the masses and work against the interests of the insurgents.

The government must ingratiate itself with the masses in order to counter insurgent activity effectively. The government must establish its respectability. With this objective in mind, the Afghan government has launched a vigorous campaign to persuade the public that Soviet troops are in Afghanistan because of a serious threat to their country's national security and would leave whenever the Afghan government asked them to do so. For their part, the Soviet military and the civilian advisors try to create their own credibility by speaking in Persian as much as possible and by observing local customs.[18] They are not only trained in their jobs but know Islamic customs. Many of them are from the Soviet "Islamic zone" bordering Afghanistan.

Muslim tribal insurgents are unable to create liberated zones or base areas in which to carry the violence of the insurgency to its logical conclusion and establish the administrative control necessary to consolidate the movement. The insurgents cannot establish a provisional government or become recognized as a legitimate contender for power. Though the movement is alive in its operational phase and is gaining strength, it has not been able to reach the concluding phase. probably because it is essentially a minority movement led and organized by tribesmen and the Shiite religious minority, rather than by a national organization like the Khalq-Parcham.

One basic human demand is to have some degree of order and predictability. Afghanistan is in a state of political and economic chaos. This condition is intensified by the machinations of the insurgents. The disruptive influence of insurgent attacks and punitive action taken by the government leave large segments of the people confused and disappointed.

The insurgency movement in Afghanistan is past the incipient phase and is going through the organizing and the operational phases. As yet, it has been unable to reach the concluding phase for six reasons.

First, there is a serious lack of leadership and coordinated opposition to the government. Second, the insurgents are seriously underarmed with outdated weapons. Third, when the Soviets sent many of their "Muslim" soldiers, the insurgents lost their major purpose, to campaign against infidel rule and communist domination of their country. Fourth, the people are getting weary of disruptive activities. Fifth, the movement is tribal rather than nationwide. Finally, the insurgents have failed to gain wide international support.

IS ANOTHER VIETNAM POSSIBLE?

If the Afghans continue to resist as fiercely as history indicates they will, the Soviets may face a situation similar to that which the United States faced in Vietnam. There are, of course, striking similarities, but also crucial differences. Like the United States, the Soviet Union inherited a demoralized and poorly trained Afghan army. Like the Vietcong, the Afghan tribal insurgents show no sign of melting away before the superior firepower of Soviet weapons. The Soviet-installed government in Kabul is as discredited as the South Vietnamese government was.

The dissimilarities are that, in Vietnam, the United States had no historic ties and no common border. Afghanistan has 200-year-long ties with the Soviet Union and an 800-mile common border. Moscow has none of the supply problems that badgered Washington in Vietnam. The United States faced hostile international public opinion, serious antiwar movements and antidraft demonstrations at home. The Soviets have no problem containing public opinion at home for their involvement in Afghanistan and there is only questionable international pressure. Ultimately, the United States withdrew from Vietnam. The Soviets are learning a lesson from the United States' involvement in Vietnam in terms of loss of manpower, money, and prestige.

Three successive governments since 1978 and the Daud administration before them have attempted to introduce many socioeconomic reforms, such as secularization, emancipation of women, and economic development, on an unwilling traditional society. Secular modernization programs are unacceptable to tradition-bound Muslim tribesmen and mullahs. On top of that the Khalq-Parcham are known throughout Afghanistan as being rigidly Marxist. This is equally unacceptable to the Afghan people, and harbors continued resistance and insurgency. The Muslim insurgency is a reaction to attempts to modernize a primitive society.

In spite of Soviet assistance and modern weaponry, Afghan military forces have not been able to put down the insurgency of tribesmen armed with outdated rifles. Despite all odds, the tribal insurgency has survived and grown in strength. Moscow is now stuck with a military adventure that seems to be not only faltering in its main objectives, but also undermining Soviet world standing.

MILITARY SITUATION

In the face of the mighty Soviet army, one of the world's largest and most powerful, the prospects of a poorly armed insurgent movement seemed initially hopeless. Yet after six years of Soviet occupation, the military situation in Afghanistan remains at a virtual impasse. The limited contingent the Kremlin dispatched has not been enough to suppress Afghan resistance. Although the Afghans are not likely ever to be strong enough to expel the invader, the Soviets slowly and steadily have been compelled to increase their forces and firepower. They must continually reevaluate their tactics just to maintain their position.

The Soviets and their Kabul allies are able to exercise effective control over only a small fraction of Afghanistan. Except for sweep operations, they rarely venture away from their bases, parts of the cities, and the major highways. At night, even these are not safe. Most of the country's rural areas remain beyond Soviet and government control. The Afghan resistance fighters are able to move throughout the country and exercise almost full authority over wide areas. In some places they effectively govern, collect taxes, and run schools.

Soviet and government efforts to establish control over the major cities and towns have met with limited success. Maintaining security in Kabul is a priority for the government, but the city has increasingly been subject to resistance. Security in the capital deteriorated sharply in late 1984, when the Mujahidin carried out a number of rocket attacks. Significant areas of Herat and Qandahar, the second and third largest cities, are under resistance control, and their populations have dwindled due to Soviet and government bombardment.

RESISTANCE CAPABILITIES

The Mujahidin have been increasingly effective throughout the six years since the invasion. Their armament has improved from traditional homemade rifles to nearly the full range of Soviet weaponry, much of it captured or handed over by deserters from the Afghan army. They are now capable of countering Soviet or government aircraft with antiaircraft guns and surface-to-air missiles. Although there are continued reports of disputes and even fighting between resistance groups, there also have been signs of increasing operational cooperation.

Despite extreme hardship and suffering, there is no sign that the resistance is losing the support of the overwhelming majority of

Afghans. The Afghan people have provided the Mujahidin with food, shelter, and recruits. Many still working for the government or army have supplied equipment, access, and inside information.

The inability to maintain an effective Afghan military has been one of the most significant problems for the Kabul government and the Soviets. The Afghan army, which had 90,000 men before 1979, has been reduced largely through desertion to 35,000–40,000. Regular conscription is supplemented by roving press gangs. The draft age has been lowered almost every year and recently dropped to 16. Such efforts themselves result in increased desertion rates. In 1983, when the term of service was lengthened, previously discharged veterans were subjected to additional service. The continued deterioration of the Afghan army has necessitated a greater reliance on Moscow's troops.

SOVIET FORCES

The Soviet Union now has about 115,000 troops in Afghanistan. They are supported by 30,000–35,000 troops stationed in the central Asian region of the Soviet Union. Theft and assault, abuse of alcohol and drugs, black marketeering, poor discipline, disease, and supply shortage caused partly by Mujahidin interdiction have reduced effectiveness. The Soviet army derives some benefit in training and equipment testing from its first real combat since World War II.

Despite low morale and 20,000–25,000 casualties (about one-third of these were killed), Moscow appears determined to remain in Afghanistan. At the same time, there has been no indication that the Soviet Union is ready to significantly expand its forces. Their current strength is enough to thwart any resistance attempt to dislodge them.

Because they cannot depend on Afghan government troops, the Soviets have been forced to play an expanded role and display a new aggressiveness, as evidenced by their major operation in spring 1984 in the Panjsher Valley. In their seventh attempt since the invasion to take this strategic valley, they resorted to high-altitude saturation bombing by aircraft based in the Soviet Union and committed many thousands of troops. While they were able to reestablish control in the lower valley, they failed to eliminate the local resistance group and its leader, Ahmad Shah Masood. Masood had used a truce offered by the Soviets in 1983 to consolidate his forces and carry out operations outside the valley. Repeated government claims that Masood was killed proved false. The Soviets have pursued a scorched-earth policy in the Panjsher Valley, destroying most of the crops and irrigation networks.

HUMAN RIGHTS IN AFGHANISTAN

The human-rights situation in Afghanistan, one of the worst in the world, results from the Soviet invasion and continuing occupation of that country, where the Soviets are trying to impose an alien political and social system. For centuries, the fiercely independent and devout Muslim people of Afghanistan have resisted foreign domination. Following the April 1978 Marxist-Leninist coup establishing the Democratic Republic of Afghanistan, opposition to the government quickly developed and spread throughout the country. Attempting to prevent the fall of a communist regime on its borders, the Soviets invaded Afghanistan and executed the Marxist prime minister.

After six years of increasingly brutal occupation, popular resistance forces dominate the countryside, keeping Soviet troops at a military stalemate. The Soviets and their Afghan allies, including the Afghan secret police or KHAD (modeled after the KGB), hope to break the will of the Mujahidin and their civilian supporters. Using ruthless terror and repression, they hope to create a totalitarian, communist state in Afghanistan.

Soviet brutality toward civilians has included retaliatory killings; torture and maiming of political prisoners; and destruction of food, livestock, and property. Reports of arbitrary killings of Afghan civilians by Soviet troops have mounted steadily since 1979. For every guerrilla attack on Soviet forces, the Soviets automatically retaliate against villages suspected of supporting the resistance movement. In many such retaliatory strikes, evidence points strongly to indiscriminate killing of innocent women, children, and old men. Another Soviet military tactic involves dropping antipersonnel mines (often disguised as household objects or toys) aimed at wounding Afghan civilians, in order to terrorize and demoralize Mujahidin supporters. Still another tactic is the destruction of farms and crops in areas supporting the Mujahidin, forcing the population to flee. The Soviet military also has carpet bombed entire villages. As a result of the bombing, killing, and widespread destruction caused by these tactics, more than 3.5 million Afghans (20 percent of the prewar population) have fled to Pakistan and Iran, with hundreds of thousands of others crowding into Afghan cities, especially Kabul, to avoid the war.

The Karmal regime has executed political prisoners following public show trials. Torture, both physical and psychological, is widely used by the KHAD to obtain information, inflict punishment on prisoners, and intimidate the population. Survivors of Afghan prisons speak of electric shock, beatings, extraction of fingernails, and other barbarous acts, often with Soviet advisers present. A number of sources, including Amnesty International, have identified at least eight KHAD detention centers where torture takes place.

Afghans in government-controlled areas live in fear of arbitrary arrest or detention by the secret police. Warrants are not used and charges may not be filed for months, if ever. Following lengthy imprisonment, political prisoners usually are tried in secret, with the KHAD playing a major role. If a trial takes place, prisoners are presumed guilty at the outset, are denied counsel, and are seldom allowed to cross-examine the prosecution's witnesses. Defendants are not often acquitted.

There is no freedom of speech, press, assembly, or movement in areas controlled by the Karmal government and the Soviets. The media are owned and controlled by the government.

Restrictions on Afghan citizens are enforced primarily by the KHAD, which numbers about 20,000 and has a pervasive influence in government-controlled areas of Afghanistan. In addition to its other activities, the KHAD has a large network of informers. As a result, Afghans are guarded in their speech so as not to be accused of anti-government remarks. The military has lost many of its troops to desertion, leading the KHAD to use press gangs in bazaars and in house-to-house sweeps to obtain conscripts.

Domestic travel has become extremely difficult due to damage to roads and bridges, and fighting between Soviet forces and the Mujahidin. Foreign travel is strictly controlled, limited to persons the government believes will return. Despite these hardships, at least one out of five Afghans has fled home to escape Soviet occupation.

Soviet forces in Afghanistan maintain the totalitarian state. Membership in the PDPA, the only legal party, is mandatory for any meaningful position in public or professional life. There has been no legislature or election of any kind since 1978. Instead, the government rules by edict.

Soviet counterinsurgency tactics have hit civilians as well as the Mujahidin. For a time, Soviet planes were dropping antipersonnel mines disguised as toys, watches, and other objects that Afghan children or refugees would pick up. These mines were designed to maim rather than kill. Frequent savage reprisals against villagers suspected of aiding the Mujahidin further alienate the population against the Soviets and the government.

Evidence indicates that Soviet troops and their Afghan proxies have used lethal chemical and toxic weapons in Afghanistan. Attacks with such weapons on the Mujahidin were reported as early as six months before the Soviet invasion and continued through 1982. Reports of chemicals and toxins used include mycotoxins (poisons derived from natural biological sources), nerve gases, incapacitants, blister agents, carbon monoxide, and nonlethal gases delivered by a variety of means. Typical targets are Mujahidin hiding in tunnels or in inaccessible mountain redoubts. Although there have been no confirmed incidents since 1982, there are indications that chemical agents are still being used.

POLITICAL SITUATION

Tainted by Soviet sponsorship, the Karmal government has been no more successful at winning popular political backing than its predecessor, in spite of major efforts to broaden support. Since its inception, the government has adopted a much more moderate approach to social and economic change. When it eventually reintroduced land reform, it included exemptions for military, religious, and tribal leaders who support the government. The government has attempted a reconciliation with religious leaders. Karmal has even tried to portray himself as supporting Islam. Public skepticism and religious opposition are fueled by government efforts to control Islam such as the removal of 20 of Kabul's most prominent mullahs from their mosques in late 1984.

In December 1980, the government, with much fanfare, announced the formation of the National Fatherland Front (NFF). Made up of tribal and religious leaders and representatives of PDPA-backed unions and social organizations, the NFF was designed to extend party influence. Efforts to persuade nonparty members to join have not been successful. Aside from token participation in government-sponsored conferences, most religious, tribal, and community leaders will have nothing to do with the Marxist government. Many, instead, are active in the resistance.

The formation of the NFF was one of a series of Soviet-sponsored efforts to offset the devastating internecine conflict between the Parcham and Khalq factions of PDPA. The Soviets insisted on including Khalqis in the cabinet and have attempted to rein in Parcham moves to extract vengeance for the oppression they felt when the Khalq was in power. These moves have borne little fruit. Continued interparty strife has resulted in assassinations of members of both factions.

Repression also affects those outside the government. The Soviet-directed KHAD and the Afghan army reportedly are responsible for torture, executions, and human-rights violations of every description.

Moreover, the Kabul government has given shelter and assistance to international terrorists. When members of the al-Zulfiqar movement hijacked a Pakistan International Airlines jet and were welcomed in Kabul, international sanctions were imposed which bar Afghanistan's Ariana Airlines from landing in many countries.

ECONOMY

Economic conditions in Afghanistan, which has little arable land, limited natural resources, and only a rudimentary infrastructure, have always been precarious. Since the invasion, conditions have deteriorated. Agricultural production has dropped, causing food shortages in some areas and, in the cities, increased food grain imports from the Soviet Union. Cultivators have fled the fighting, the draft, or joined the resistance. For those who remain, it is often too risky to go out into the fields. Because of input disruptions and Mujahidin sabotage (since most industry is state owned), industrial production also has declined.

The only growing parts of the economy are those linked to the Soviet Union. The Soviets encourage such ties. Natural gas, Afghanistan's major natural resource and export item, is piped directly from the ground to the Soviet Union and sold below world market prices. No distribution pipelines run to Afghan cities. A major copper mining project is planned to produce copper for direct export to the Soviet Union. Outside the thriving black market, statistics show little trade with the noncommunist world.

REFUGEES

Partly as a result of increased military activity, refugees continue to leave Afghanistan, although at a reduced rate from earlier periods. Pakistan remains host to an estimated 2.5 million refugees, the largest refugee population in the world. An estimated 800,000 more Afghan refugees have fled to Iran.

In addition to those who have left Afghanistan, an undetermined number of people have been displaced within the country. Fighting and destruction have driven people into urban areas. Since the Soviet invasion, Kabul's population has doubled, despite the exodus of large numbers of urbanites.

Despite the heavy burden they have placed on Pakistan, the refugees have been welcomed. A good rapport continues between them and their hosts. Pakistani assistance to the refugees includes cash allowances and payment of relief administration costs.

INTERNATIONAL REACTIONS

The world reacted with shock and horror at the Soviet invasion and the brutal war. The United States denounced the invasion and imposed a number of sanctions against the Soviet Union. Participants in the Organization of Islamic Conference, demanding a pullout, voted to suspend Afghanistan's membership and called on its members not to recognize the Karmal government. The nonaligned movement called for the withdrawal of all foreign forces.

From the very beginning, the Kabul government has maintained that the presence of Soviet troops is a bilateral matter. A troop withdrawal could be considered only after outside interference (the resistance) ceased. The DRA proposals call for international guarantees of noninterference as part of any settlement.

UNITED NATIONS EFFORTS

In January 1980, the United Nations General Assembly condemned the Soviet invasion by an overwhelming vote. In November of that year, the General Assembly passed a resolution calling for the withdrawal of foreign forces from Afghanistan and calling on the secretary-general to seek a negotiated solution. A similar resolution has passed each succeeding year. In 1984, the resolution, sponsored by Pakistan and 46 other nonaligned states, won a record vote of 119 to 20 with 14 abstentions. This wide margin reflects the continuing censure of the Soviet occupation by the world.

The major elements of the General Assembly resolution provide the basis for a settlement and United Nations-sponsored negotiations. These are: the complete withdrawal of all foreign troops, the restoration of the independent and nonaligned status of Afghanistan, self-determination for the Afghan people, and the return of refugees with safety and honor.

In 1981, negotiating efforts were begun by the secretary-general's personal representative, Javier Perez de Cuellar. When de Cuellar himself became secretary-general, he appointed undersecretary-general Diego Cordovez to the position.

In June 1982, after a series of separate consultations, Pakistan and the Kabul government sent delegations to Geneva for indirect talks led by Cordovez. Iran, which strongly condemns Soviet occupation, did not directly participate because "the real representatives of Afghanistan" were not invited. Iran did agree to be kept officially informed. Although there were reports of progress and flexibility, no agreement was reached. The Geneva talks reconvened in April and June 1983 and in August 1984 without progress.

OUTLOOK

The insurgency that began against the Afghan Marxist regime, and intensified after Russian occupation, is not likely to end even with Soviet withdrawal. A national government is needed, involving all political parties and groups in a nationwide election for a constitutional assembly which would then determine the form and character of the new Afghanistan political structure. After Soviet withdrawal, if Afghan Marxists continue to rule the country, in all probability

Muslim insurgents will continue their resistance against the government. If the tribal insurgents should be successful, it would be the first time Muslim insurgents have repelled communist invaders.

The outlook for the future of Afghanistan is grim. Fighting is expected to continue for the forseeable future. The Soviets seem intent on a long-term strategy based on maintaining the regime in Kabul, wearing down the resistance, and the Sovietization of the Afghan government, economy, society, and people. The Mujahidin appear willing to pay the heavy cost of continuing their struggle, and say they will not give up. Much will depend on the outside world's attitude toward the conflict.

NOTES

1. Thomas L. Brewer, *American Foreign Policy: A Contemporary Introduction* (Englewood Cliffs, N.J.: Prentice-Hall, 1980), p. 19.

2. Rouhollah K. Ramazani, *The Northern Tier: Afghanistan, Iran, and Turkey* (Princeton, N.J.: D. Van Nostrand, 1966), p. 26.

3. Louis Dupree, *Afghanistan* (Princeton, N.J.: Princeton University Press, 1973), p. 110.

4. M. N. Roy, "Democracy and Nationalism in Asia," *Pacific Affairs* 25, no. 2 (June 1952): 140.

5. R. T. Klass, "The Tragedy of Afghanistan," *Asian Affairs: An American Review* 7, no. 1 (September–October 1979): 5.

6. Hannah Negaran, "Afghanistan: A Marxist Regime in a Muslim Society," *Current History* 76, no. 446 (April 1979): 173.

7. *Economist* (London), June 14–20, 1980.

8. Ibid.

9. Alvin Z. Rubinstein, "The Soviet Union in the Middle East," *Current History* 77, no. 450 (October 1979): 106–09.

10. Joel Larus, "The End of Naval Detente in the Indian Ocean," *World Today* 36, no. 4 (April 1980): 126–32.

11. *Encyclopaedia Britannica, Book of the Year, 1979* (Chicago: Encyclopaedia Britannica, 1979), p. 99.

12. Insurgency is defined as a condition of revolt against a government that does not reach the proportion of an organized revolution.

13. Klass, "The Tragedy of Afghanistan," p. 7, and *U.S. News and World Report*, February 4, 1980.

14. *Time*, February 4, 1980.

15. *Washington Post*, February 14, 1980, *Durham Morning Herald*, February 16, 1980, and *Newsweek*, February 25, 1980.

16. Dupree, *Afghanistan*, p. 659.

17. *Time*, February 4, 1980.

18. *Washington Post*, January 25, 1980.

6

The Arab–Israeli Conflict and the Oil Weapon

Because of the age-old animosity between Arabs and Israelis and the incessant military confrontations between them, the Middle East is almost always the focus of world attention. In this chapter, we will describe, first, the nature of the Arab-Israeli conflict and the parties involved and, second, the use of Arab oil as a weapon of diplomacy in the settlement of the Arab-Israeli dispute. It is critical to recognize that the protractedness of the conflict has worked itself to change both the nature and the parties of the dispute, thus complicating the issues.

ROOTS OF THE CONFLICT

The roots of the Arab-Israeli conflict go back to 1897 when a group of Jews, following the initiative of Theodor Herzl, passed a resolution favoring a home for Jews in Palestine and created the World Zionist Organization to carry it out. Since the dispersion religious Jews have cherished the messianic hope of one day returning to the Holy Land. Herzl's was essentially a spiritual sentiment altogether unconnected with political ambition.

At this time the Arab world was passing through some of its most difficult days. The French capture of Algiers in 1830 paved the way for Western domination of the Arab world. At the same time an Arab revival—a renewed sense of language and cultural pride and a desire for self-government—was emerging. These sentiments found little sympathy with Ottoman authorities, with their predilection for centralism, and were adroitly exploited by their enemies.

In 1914, Britain, fearing a Turco-German offensive in the Middle East, asked Hussein, the grand sharif of Mecca, for an Anglo-Arab alliance against Turkey. Though eager for independence, the Arabs were held back by suspicion of European designs and an aversion to fighting against Muslims. Later, the sharif was tempted to accept the British offer on the basis of the alluring prospect of complete independence of all Arab lands then under Turkish rule, including the whole of Iraq and Syria, of which Palestine formed an integral part. The formal acceptance of Arab conditions, with only minor modifications, was signified in a note from Henry McMahon, the British high commissioner in Egypt, to the sharif on October 24, 1915.

On the strength of this pledge, the Arab revolt was proclaimed on June 5, 1916, decisively tilting the military balance in favor of the Allies. Britain, however, had already betrayed Arab trust. According to the Sykes-Picot Agreement signed with France in May 1916, Britain and France were to divide Iraq and Syria between themselves, while Palestine was to be detached from Syria and put under an unspecified international rule. Late in 1917, when Hussein heard of the secret agreement and demanded an explanation, an ingeniously worded telegram from Arthur Balfour, the British foreign secretary, deceived him into believing that no such agreement existed, and the Arabs continued to shed their blood for the Allies.

With the installation of the mandatory government in 1920, under a Jewish high commissioner, the gates were thrown open to tens of thousands of European immigrants. From the start, Jewish colonization was aimed at establishing Zionist supremacy through systematic dispossession of Arabs and the creation of a closed economy. With the help of the Jewish Agency, the colonists were allowed to buy, often for a pittance, lands formerly owned by Syrian and Lebanese landowners now cut off from their property by the new frontiers. Once bought, these lands became legally Jewish and could never be repurchased by their rightful owners at any price. Entire Arab villages were wiped out this way.

Finding their appeals for a democratic parliamentary government and a ban on further immigration and land sales falling on deaf ears, and growing desperate by the inexorable advance of Zionist colonists, the Arabs reacted the same way all people living under colonial rule have reacted in similar circumstances. The first Arab outbreaks occurred as early as 1920 and continued for a year, when Winston Churchill, the colonial secretary, assured them on June 3, 1921 that no Jewish state was contemplated in Palestine and that immigration would only be allowed when compatible with the "economic absorptive capacity of the country."

The number of European settlers kept swelling. In the first ten years of the mandate Jewish immigration tripled from 10,000 to 30,000 a year, and by 1935 had soared to 62,000. An almost continuous joint Zionist-British action was required to quell the disturbances that followed the inevitable dispossession.

In 1939, in view of growing German and Italian propaganda in Arab countries, and the threat of war, the British government issued the White Paper reversing its decision on partition and proposed a single binational state. This state would come into being after a transition period of ten years. Seventy-five thousand Jews were to be allowed to enter Palestine over five years, with further immigration dependent on mutual Arab-Jewish consent. Britain succeeded in bringing about an uneasy truce in Palestine, and prevented a concerted Arab revolt.

World War II provided a unique opportunity for the Zionists to equip themselves with arms partly obtained from Allied troops and partly smuggled in through organized international trafficking, and to acquire large-scale military training in the British army.

Subjected to growing American pressure and at odds with both Arabs and Jews, Britain decided to wash its hands of the problem by submitting it to the United Nations on April 2, 1947. The United Nations special committee report was not unanimous: a minority plan presented by India, Iran, and Yugoslavia recommended a single federated state with restricted immigration, while the majority plan proposed partition, with the 30 percent of Jews getting 56 percent of the whole area, while the Arab majority was to get the barren hills of Samaria and the wilderness of Judea, and was denied even the Nejev, which was awarded to them by the royal commission. Jerusalem was to remain under international control (the Jewish sector was provided with 56 percent of the territory, the Arabs were given 42 percent, and the remaining 2 percent in Jerusalem were international). The Zionists prevailed upon the U.S. government to take the lead in sponsoring the creation of a Jewish state. On November 29,

1947, after frenzied lobbying, the United Nations General Assembly, by a vote of 33 to 13 with 10 abstentions, passed the majority partition resolution.

THE BIRTH OF ISRAEL

On May 14, 1948 the British mandate was formally terminated. The same day the state of Israel was proclaimed and within hours was granted de facto recognition by the United States. Soon afterward, the Arab armies of neighboring countries entered Palestine in defense of Arab rights. The main interests of the Arabs lay not necessarily in the immediate destruction of the new Jewish state (contrary to the accepted belief), but in the internal Arab struggle for power and position in the region. When fighting broke out between Israel and the Arabs, Arab nations not having a frontier contiguous with Palestine did not send troops or take an active part.[1] Egypt, Jordan, Iraq, and Syria joined the Palestinian war and fought against the newly created Israel, but the Arabs lost the war.

In 1956, when Britain and France attacked Egypt on the issue of nationalization of the Suez Canal, Israel joined in, making it the second round of the Arab-Israeli war. Many Arab nations ordered mobilization and called on all Arabs to oppose the attack. The war ended in an Egyptian military loss but a political victory—the Suez Canal was nationalized.

The third war between the Arabs and Israelis broke out in June 1967. This time the Arabs called for the annihilation of Israel. But the six-day war again ended in Arab defeat and loss of significant territories.

After the immediate shock of the 1967 defeat, Egyptian President Nasser assumed that what had been lost in war could be restored only by war. Nasser's successor, Anwar Sadat, also concluded that there was no way to break the stalemate other than war. In an April 1973 interview with *Newsweek*, Sadat declared that "everything in this country is now being mobilized in earnest for a resumption of the battle."[2]

The purpose of the fourth Arab-Israeli war (October 1973) was not to destroy Israel but to regain the occupied territory. Arab attitudes toward Israel have changed since the 1967 war. Until then the Arabs had demanded the complete dismantling of Israel. They mod-

ified their stand against Israel and engineered the passing of United
Nations Resolution 242 which called for an Israeli withdrawal from
all occupied territories. The Arabs reaffirmed their stand not to
"throw Israel into the sea" during the war when the resolution imposing the oil embargo was adopted in Kuwait. This Arab willingness
to reconcile with the Jewish state grew not because of a love for
Israel but because of their failure to regain territories by attrition or
diplomacy.

Except for the first, oil figured significantly in all Arab-Israeli
wars. In 1948, Arab oil production was negligible. Iran was the major
Middle East producer. Britain still had complete control of the oil
supply passing through the Suez Canal. In 1956, oil shipments via the
canal to Western countries were blocked. In 1967, the canal was
closed indefinitely.

In 1973, Egypt and Syria launched full-scale military attacks
against Israeli occupation forces. The Arab oil weapon fully came
into play. There was a five-month embargo on oil shipments to the
United States and the Netherlands. A partial embargo was in effect
against other nations.

Although the decision would be much more serious than in 1973,
most observers believe that a renewed Arab-Israeli war would precipitate a more severe oil embargo. It is also axiomatic that the United
States would not permit Israel to be destroyed in a war. By the same
token, Soviet support for the Arab cause cannot be assured. The
Soviet Union backed out of its support of the Arab war effort in
1967. Again in 1973 it did not fulfill its announced intention to send
volunteers to the Middle East as soon as the United States put its
strategic forces on alert.

The two superpowers may come to terms in the new economic
game now in progress in the Middle East. All nations are interested
in securing oil supplies at affordable prices. By imposing an embargo
the producing and exporting Arab nations turned the Middle East
crisis into a global trauma. Even the communist nations and many
Third World countries—the main supporters of the oil price increase—
are feeling the pinch of the energy crisis.

Diplomacy has had a chance to settle the Middle East crisis, and
its value is not completely lost. It can still succeed, provided the
Palestinians refrain from calling for Israel's destruction and settle for
a homeland on the West Bank of the Jordan River and the Gaza Strip.
A growing number of specialists maintain that unless an agreement is

reached soon, war may come to the Middle East during the next few years. One prominent pundit maintains that quiet diplomacy may be an effective means of settling the Arab-Israeli conflict.[3] Quiet diplomacy is negotiation from a position of helplessness. The United States' helplessness emanates from its failure to stop oil price increases and guarantee the flow of oil to the industrialized world.

The situation in the Middle East has been described as follows:

> With the Palestinians playing a newly pivotal role in intra-Arab politics and with the Arab world as a whole seemingly convinced that oil will turn the tables on Israel, Rabat and its aftermath seem to have made new war much more likely. It could break out in a number of ways, ranging from Arab attacks in response to Israeli rejections of demands for new withdrawal from the occupied territories to an Israeli preemptive war triggered by circumstances such as a new Arab "war of attrition," evidence of a coming Arab attack, or simply a feeling that a blow must be struck before the tables are hopelessly turned.[4]

In this situation of Arab belligerency Israel would go to war against the Arabs, but it cannot without diplomatic support. That support will have to come from the United States. Will the United States give its blessing to Israeli war efforts? The United States may either help Israel fight or it may fight the Israeli war itself. In the past the United States has helped Israel fight its wars through arms aid and diplomatic support, and is likely to do so in the future.

Assuming that the United States is able to persuade Israel not to launch a preemptive attack against the Arabs, the next question is whether the Arabs and Palestinians will stand idle while their territories remain under Israeli occupation. The answer is no.

> Processes of modernization throughout the Arab world have considerably increased Arab abilities to fight a sustained war. And the decisive change in the economic balance between Israel and the Arab states brought about by the ... increase in oil prices ... has removed all resource constraints from Arab armament efforts.[5]

PSYCHOLOGICAL UNDERPINNING

The reason oil again dominates the news is not the threat of another immediate embargo but the almost continuous price hikes of the past decade, and the recent downward (and probably temporary) turn of price. If an Arab-Israeli settlement is reached, will the price of oil decline? If it does, the specter of an international financial crisis would disappear.

The Arab oil leaders have never promised a specific price reduction or concessions in return for a settlement of the Arab-Israeli conflict. If a settlement is reached through the peace efforts of the United States or other outsiders, will the Arabs feel obligated to lower prices in gratitude for the help? The answer depends on whether the radical Arabs will abide by any settlement short of the complete destruction of Israel. Unless the Arabs have a prior legal or moral obligation, they will raise the issue of their sovereign rights to decide the price of oil as they see fit in association with other oil-exporting nations. Historically speaking, oil producers and exporters have not always fulfilled their legal obligations. They have violated many agreements made with oil companies. Thus, the Arab-Israeli conflict will remain a potential danger to the world oil supply and international peace and order.

The Arab-Israeli conflict has created a political climate in the Third World in support of oil price increases. Essentially, the price increases were a revolutionary attempt to seize control of the resource and radically redistribute the wealth. The political and strategic implications of the oil weapon must be seen in light of the role of the oil weapon in the struggle for political hegemony among various preindustrialized nations. The London-based International Institute for Strategic Studies examined this changing situation, concluding:

> This was the first time that major industrial states had to bow to pressure from pre-industrial ones. . . . The victory upset the hierarchies of power long enjoyed, or resented, according to one's station, and opened up prospects of quite new political balances. By the same token, it was by far the biggest extension of the world's effective political arena since the Chinese Revolution.[6]

The Middle East situation, with regard to the Arab-Israeli conflict, has changed little, except that under the Camp David agreement Egypt got back its land from Israel. The question now is whether the Arabs can use their petrodiplomacy bargaining power in the future. The task of OAPEC is to persuade targeted states to change their policy toward Israel and force the United States to bring Israel to change its policy. The present oil glut would make this task more difficult should it continue.

THE ARAB OIL WEAPON

The oil-exporting Arab countries had practically nothing to do with the development of the oil industry. The hidden treasure just happened to be in their territories. Today the Arabs have the greatest impact on the international oil industry. As a group, these OAPEC nations are a very important world oil supplier. Many of them are militant, which poses a serious threat to the oil flow to other nations. Without oil, many countries would be significantly less important. Even with oil, these nations have no real impact on the world stage when considered individually.

No oil-exporting country at present has a dominant position in the international oil market, although Saudi Arabia's oil is almost inexhaustible. It is because of this huge oil reserve that Saudi Arabia carries the greatest weight in OAPEC. Not all OAPEC nations can follow the supply-demand production ratio over a long period because of their greater need of oil revenue. Only Saudi Arabia, Kuwait, and Libya are capable of increasing or lowering production to keep supply on a reasonable balance with demand. Added to this is the huge accumulation of petrodollars by Saudi Arabia, Kuwait, and Libya. This allows them to control OAPEC's and even OPEC's success or failure.

The driving force for change is oil, which is not only shaping the structure and character of Arab society but is radically changing the economic and political significance of the region.

The concept of using oil as a diplomatic weapon against the West and Israel is as old as the Arab-Israeli conflict itself. Anticipating Western support of Israel, the Arab League passed a set of resolutions in June 1946, one of which called for denial of oil to the West.[7] When fighting broke out in May 1948 in Palestine, the embargo was not implemented, largely because of opposition by Saudi Arabia, "which believed that a commercial oil operation should be divorced from political considerations."[8] Considering that, at that time, all Arab oil was being produced by the international oil companies, the Arab League realized that its members had no voice in deciding the level of production, price, or export of the commodity.

Nevertheless, this premature attempt to use the oil instrument as a diplomatic weapon to settle the Arab-Israeli conflict met with some success. George Lenczowski notes:

> In a gesture of defiance toward Israel and out of solidarity with other Arab states, Iraq stopped the movement of oil by pipeline to the Israeli-held Haifa terminal and caused construction of the parallel line between Kirkuk and Haifa to cease . . . and . . . boycott measures against Israel by the Arab League gradually affected the transactions of a number of oil companies with Israel.[9]

Of the four Arab-Israeli wars, oil figured significantly in all but that of 1948. In 1956, when Egypt nationalized the Suez Canal, Britain and France declared war against Egypt. This turned out to be the second Arab-Israeli war when Israel joined Britain and France. The Arab countries not only blocked oil to Israel, but to the Mediterranean via the Suez Canal. The pipelines from Iraq and Saudi Arabia were closed. Some pipelines were blown up by nationalist elements in retaliation for the tripartite attack on Egypt. The situation arising from the closing of the Suez Canal and the embargo of Arab oil to Britain and France produced serious economic consequences. Oil embargoes increased oil prices in Britain and France. The oil companies had to obtain oil from the United States, which shipped their Middle Eastern oil products via the Cape of Good Hope. This entailed more costs. As a result oil rationing was introduced in Western Europe.

During the Suez war, the United States found itself in "alliance" with the Soviet Union. They both played a major role in calling upon Britain, France, and Israel to cease fire and withdraw their forces. Directly or indirectly, the two superpowers urged Egypt and other Arab countries to abandon the embargo on oil shipments to their adversaries. Consequently, Arab attempts to use their control over oil as a diplomatic weapon for a long period failed without settling the Arab-Israeli conflict. The oil embargo which lasted for six months had no lasting impact on Western European economics.

In 1967, when the third round of the Arab-Israeli war began, Mideast oil was even more important to most of the industrialized world. Oil has almost replaced coal since 1965. When the war broke out, Nasser announced that Britain and the United States had joined Israel in attacking the Arabs.[10] He also said that the U.S. Sixth Fleet helped Israel attack Egyptian airports and military bases. These accusations were not true, but at that time the Arabs took them seriously. Notices were sent to oil companies to cease exporting oil to the countries, such as Britain and the United States, blacklisted by the Arabs. Germany was added to the list because it sold gas masks to Israel. For the first time Arab oil-producing nations completely shut down oil production. In further retaliation, in Saudi Arabia and Kuwait, "bands of saboteurs assembled with explosives, ready to destroy the big companies' installations once and for all."[11] Before the oil fields could be blown up Arab troops moved in to occupy them. Soldiers were ordered to stop the oil flow, and their presence prevented sabotage. The pattern of the 1956 oil embargo was repeated by the closing of the Suez Canal, this time indefinitely.

The European economy was not as badly hurt as in 1956, because oil was imported from Libya, Algeria, and Venezuela. The United States also exported oil to European countries to help meet the crisis.

The oil embargo was short. Only two weeks later oil flowed again. There were several reasons for its failure. First, because of inaccurate propaganda, the Arabs decided to use their oil weapon as the only alternative to inaction. Second, within a few days all of them found that they did not have the financial strength to carry on the oil embargo without getting money from the oil companies. "Saudi Arabia was the first to feel the pinch acutely.... King Faisal was informed by his finance minister that there was no more money in the till, and that for once Aramco was unable to help."[12] Finally, participants in the Arab summit conference, held in Khartoum shortly after the war, decided that oil should be used "positively" as a political weapon, the implication being that their approach was based on false propaganda and other wrong notions. A statement in this connection made by the undersecretary of the Kuwait Ministry of Foreign Affairs said:

> Kuwait has always subscribed to the notion that oil should be used as a weapon in the confrontation with Israel, and has, however, applied this notion ever since it established the Kuwaiti Fund for Arab Economic Development, which strengthens the Arab economy by financing development projects that have helped and continue to help Arab steadfastness ... it is not in anyone's interest to use oil negatively.[13]

The use of oil as a weapon in 1967 had been badly handled. "Injudiciously used, the oil weapon loses much if not all of its importance and effectiveness," said Yamani.[14] On the Arab maneuvering of oil, he said further "if we do not use it properly, we are behaving like someone who fired a bullet into the air, missing the enemy and allowing it to rebound on himself."[15]

The oil shots failed because, first, the United States was not hurt by them. On the contrary, the embargo enabled the international oil companies to make handsome profits through increased production of U.S., Venezuelan, and North African oil. Second, despite the closing of the Suez Canal, the oil companies were able to supply oil to their customers in Europe without much difficulty. Third, no quota ceilings were imposed by the Arab governments, which encouraged overlifting oil from Mediterranean and other ports.

"Oil is a formidable weapon if it is used properly," said King Hussein, the ruler of the oilless kingdom of Jordan.[16] The proper and positive use of oil as a diplomatic weapon can be made if enough funds are available to sustain the long and arduous task of implementing an embargo and helping the have-nots in the Arab world.

Following the June war of 1967 the whole atmosphere of inter-Arab politics changed. Although the Arabs were not motivated by international considerations, they were beginning to give serious thought to Arab unity. The message of military defeat reached equally the belligerent and nonbelligerent. "There could hardly be a competition for prestige when there was no prestige remaining."[17] Self-interest and rivalry did not deter concrete action aimed at unity and the sharing of wealth as before. The Arab oil countries have been making financial contributions to oilless and oil-short Arabs since the 1967 war.

The Arab countries involved directly in the conflict with Israel are Egypt, Syria, and Jordan. These are not big oil-producing nations, although some oil is produced in Egypt and Syria. These countries call on the big oil-producing Arab nations to use their oil as a weapon against Israel in the name of Arab solidarity. For example, "in 1967 Saudi Arabia cut the flow of oil involuntarily, under pressure by Nasser, and therefore did not enforce the measure strictly and cancelled it as soon as possible."[18] Pressure or no pressure, Arab nations respond to the call of Arab unity and use their oil weapon against Israel or the West whenever the occasion demands.

As pointed out previously, the world's oil industry structure changed in the early 1970s with the disappearance of surplus production in the United States and the growing control of multinational oil companies by OPEC nations. Industrialized noncommunist countries depended on Arab oil. Oil industries in the United States stressed the significance of Arab oil, especially that of Saudi Arabia, as a most important source.

In 1972, the economic council of the Arab League made a study of the strategic use of Arab economic power vis-à-vis Israel and the energy-consuming world. The report did not call for an embargo but noted that restricting oil production would bring pressure on the consuming nations to alter their uncompromising policy toward Israel.[19] Saudi Arabia and other Arab nations gradually came under pressure to reduce their oil production.

Upon the advice of King Faisal, Sadat expelled the Soviets from Egypt so the United States could pressure Israel to withdraw to its 1967 borders. The Nixon administration failed to fulfill the commitments it had made to Faisal. As a final attempt at a peace settlement, Sadat sent his national security adviser, Hafez Ismail, to Washington to prevail upon President Nixon to make good his commitment to Faisal. At that time Israel's premier, Golda Meir, was visiting Washington. Nixon yielded to Israeli pressure by making public his intention of supplying Israel with Phantom jets. This infuriated both Sadat and Faisal.

THE 1973 OIL EMBARGO

Faisal, who had gained considerable prestige and respect among Arabs and Muslims, was very much embarrassed by the renewed U.S. desire to supply sophisticated arms and equipment to Israel. He lost face with Sadat and other Arab leaders. Without further delay he dispatched his oil minister, Yamani, to Washington in April 1973 where he linked oil and politics for the first time. Yamani told U.S. officials that it was impossible for Saudi Arabia to work against Arab interests. The U.S. officials did not take Yamani seriously, ignoring him in the belief that he was speaking for himself rather than Faisal. To remove these misgivings, Faisal told Aramco's president in Saudi Arabia that he was "not able to stand alone much longer" while pressure was building for the use of oil as a weapon.[20] Following Faisal's warning, Aramco launched a campaign in Washington, urging that the United States follow an even-handed policy in the Middle East, but without success. The Aramco official accused the powerful U.S. Jewish lobby of preventing the company from changing U.S. policy of informal partiality toward Israel.

On May 15, 1973, Algeria, Kuwait, Iraq, and Libya stopped oil production for a short time as a symbolic message to the world that Arab oil producers could and would use their oil weapon against oil-consuming nations at appropriate moments. About the same time, Libya, taking action against foreign holdings, nationalized American Bunker Hunt and took over the majority interests in Occidental and Oasis. Saudi attitude showed that nothing had really changed in its policy of moderation and friendship with the United States. Saudi Arabia's policy was to wait and see.

The oil-producing Arab nations had already increased oil revenues and production beyond what was needed for social and economic development. The Saudi kingdom had ideas for imposing the embargo selectively even before war broke out in October. On September 4, 1973, the *Christian Science Monitor* carried an article, by staff correspondent John K. Cooley, regarding an interview with Faisal and his son, Prince Saud al-Faisal, which had been published in a Lebanese paper. The Cooley article noted in part:

> ... oil is not an artillery shell, but an enormous weapon. All economic weapons need study and time for effectiveness to appear. Talk of using the oil weapon ... makes it sound as if we were threatening the whole world, while it is understood that our purpose is to bring pressure to bear on America ... but America would be the last to get hurt, because the U.S. will not depend on Arab oil before the end of the 1970s, whereas Japan and Western Europe depend on it now. What benefits are there ... from arousing the fears of the Europeans and Japanese at a time when they are showing greater sympathy for us? ... Arab policy is called upon today to persuade the American and European citizen that his interests are with the rights of the Arabs and that we do not intend to harm him, but that it is the policy of his government that is creating the confrontation [with Israel]. We must tell the American and European people that we want to defend ourselves, not harm them.[21]

When the war broke out on October 6, it looked at first as though the Arabs would not need to use the oil weapon. As the war turned against the Arabs, OAPEC oil ministers met in Kuwait on October 17 to consider the role of oil in the Arab struggle. In their attempt to bail out Egypt and Syria, which were directly engaged in the war against Israel, the Arab oil ministers, considering that the ultimate goal of the current struggle is the liberation of the Arab territories occupied by Israel in the 1967 war, and the restoration of the legitimate rights of the Palestinian people in accordance with United Nations' resolutions,

considering that the United States is the principal and foremost source of Israeli power that enabled it to continue to occupy their territories,

considering that the industrial nations have a responsibility of implementing the United Nations resolutions, and

considering that the economic situation of many Arab oil producing countries does not justify raising oil production, although they are willing to make an increase to meet the demand in those industrial nations that are committed to cooperation in the task of liberating occupied territories,

decided that each Arab oil exporting country immediately cut its oil production by a rate not less than 5% from the September production level, and further increase of 5% from each of the following months, until such time as the international community compels Israel to relinquish occupied Arab lands, and to levels that will not undermine their economies or their national Arab obligations.[22]

October 17 was a red-letter day in the history of the energy crisis. All Arab oil countries joined in the war with Israel by cutting oil production 5 percent and threatening to raise that cut monthly by 5 percent until Israel withdrew from occupied Arab territories. The countries joining in the embargo were Saudi Arabia, Kuwait, Iraq, Libya, Algeria, Abu Dhabi, Qatar, Bahrain, Egypt, and Syria. The next day, Saudi Arabia independently cut production 10 percent. The following day, Libya announced that it would raise the price of oil 28 percent, although there had been several small increases before the October war began. Iraq announced a 70 percent increase. When the oil ministers met again in Kuwait on November 4, they decided that the initial reduction in output would be 25 percent of the September level. A further cut of 5 percent would be made during each of the following months to "levels that will not undermine their economies or their national Arab obligations."[23] Meanwhile, following massive deliveries of U.S. arms to Israel, a total halt of oil to the United States and the Netherlands was announced by Saudi Arabia, Abu Dhabi, Kuwait, and Algeria. The actual reduction of oil to other industrialized nations varied between 5 and 10 percent.

While regretting the move, the Arab oil ministers said that Israel had contributed to a 5 percent reduction by bombing an oil terminal in Syria and forcing a 50 percent reduction of oil through the Trans-Arabian Pipeline. The ministers appealed to the world to "help us in our struggle" against occupation. Although the Arab countries wished to cooperate, and the oil producers were ready to supply the world with oil, the oil ministers said the time had come for condemnation of Israel's aggression.

The Arab nations' willingness to use oil as a subtle weapon of diplomacy has been attributed to their enhanced concern about Israel, and to their increased bargaining power gained from accumulated revenue:

> As long as Israel had been hemmed in within the pre-1967 boundaries, surrounded by a ring of Arab states that contained it and threatened to roll it back, countries like Saudi Arabia, Kuwait, and Iraq could feel completely safe from any Israeli threat. Whatever contribution they made to the Arab cause against Israel was made purely on the grounds of pan-Arab considerations. But as Israel overwhelmingly . . . demonstrated a . . . capacity to hurt them in a significant way . . . . Their concern with Israel . . . became an investment in their own security.[24]

Following the death of Nasser and the subsequent improvement in Saudi-Egyptian relations, Faisal demonstrated his increased ability to guide his country's foreign policy more independently:

> . . . in the past . . . Faisal feared that once he had sprung the "oil weapon," others, particularly Nasser, might be able to arrogate to themselves the right to decide when and how it was to be used.[25]

Each Arab oil-producing country administered the embargo decision reached in Kuwait in its own way. Saudi Arabia separated countries desiring oil into three categories: friendly countries, unfriendly countries, and all others.[26] The friendly countries included Britain, France, Spain, Pakistan, Malaysia, and all non-oil-producing Arab nations. The unfriendly countries were to receive no oil at all. The friendly countries were to continue to receive oil at the preembargo level. "When they had been served," said King Faisal, then whatever was left might be distributed to the rest of the world.[27]

As the Arab embargo went into effect, panic ensued in Europe and Japan. The situation was not one of absolute panic in the United

States, but the embargo caused serious concern. Both radical and conservative Arab states embargoed oil shipped to the United States, the Netherlands, and Israel. Saudi Arabia, the least radical Arab state, wanted to pressure the United States to minimize its support for Israel. Iraq, one of the most radical, believed that Europe and Japan, which were increasingly friendly toward the Arab cause, should be supplied with oil. The Arabs knew that nothing spectacular could be expected from the United States, which had a declared pro-Israeli bias for obvious domestic political reasons.

Europe and Japan, gravely threatened by their heavy dependence on Arab oil, were subjected to serious Arab diplomatic pressure or "blackmail" as news reports indicated. At times the Arabs issued ultimatums and verbal threats that a rigorous oil embargo would be imposed on Europe and Japan if their policies toward Israel were not changed. Indeed, under pressure, Britain and West Germany, both known for their long pro-Israeli biases, banned arms shipments to the combatants, including Israel. London even stopped American transport planes from landing on British territory, and Bonn protested the shipment of U.S. arms from Germany. Thus, a rift was developing in NATO, and the United States was left alone in its support of Israel.[28] The European governments denied that they had been threatened in any way by the Arabs. As the British foreign secretary said, "There has been a great deal of talk recently about submission to Arab blackmail .... The Arabs have made no demand on us and we have offered no price."[29] The fact is that while Britain and other European countries were adopting a neutral position toward the Arab-Israeli conflict, the supply of oil from the Arab world was becoming progressively more assured. As the *Sunday Times* pointed out, "Britain's oil is safe—if we behave ourselves."[30] Another important power in Europe, France, because of its special relationship with the Arabs, expected better treatment from them and indeed received an almost uninterrupted flow of oil. Meanwhile, Japan attempted to retain its neutral position by continuing to support United Nations Security Council Resolution 242, which called for Israeli withdrawal from occupied Arab lands.

How is it possible that the few desert Arab nations could seek wide compliance with their policy from the Western nations and Japan, which suddenly moved from a wasteful habit of living toward conservation of resources? To answer this critical question, let us resort to a technical description of oil consumption in the West and Japan.

In the fall of 1973 the Arab nations were producing 19.1 million b/d. The United States then consumed 17.2 million b/d and depended on Arab exports for about 10 percent of this. Europe consumed 15 million b/d and depended on Arab sources for about 65 percent of it. Japan consumed 5.2 million b/d, of which 50 percent came from Arab countries.[31]

World dependence on Arab oil increased from 34 percent in 1957 to 54 percent in 1973.[32] Let us review the impact of this dependence on prices as OAPEC put its petrodiplomacy into use. The price increase occurred in three steps during the embargo. Prior to the outbreak of the Yom Kippur war (or October war) on October 6, 1973, the Saudi government's revenue per barrel of oil was about $1.90; when the embargo was lifted on March 18, 1974, the revenue had risen to $9.25.[33]

The first price increase, on October 16, 1973, was relatively uninfluenced by the embargo since the decision had been made beforehand. The six largest oil-producing countries on the Persian Gulf—Iran, Iraq, Saudi Arabia, Kuwait, UAE, and Qatar—announced a price increase from $3.01 per barrel on October 1 to $5.12 on October 16. The next price increase by these nations came on December 23, effective January 1, when the price per barrel was raised to $11.65. The actual market price rose to $15–17 per barrel.[34] Libya and Nigeria set a new tax reference price, ranging from $14.60 to $18.75 per barrel.

Speaking for the oil producers, the shah of Iran, arguing that the posted prices of oil should be adjusted upward to reflect the market price (posted price or tax reference price is about 40 percent higher than market price), wanted "to junk the current pricing system devised by the international oil companies and base future prices on the costs of supplying alternate sources of energy."[35] Libya and Algeria wanted to push prices as high as possible. They were opposed by Saudi Arabia, which proposed a smaller increase "because it wanted to keep the politically motivated embargo separate from OPEC's pricing policy to avoid the impression that the embargo had been imposed for monetary reasons."[36] Saudi Arabia felt that a sharper price increase would debase the political message the Arabs had hoped to communicate.

The next increase occurred in January 1974 when Kuwait announced its 60 percent participation formula. The price went up by more than $1 per barrel. These increases are rational, said the oil producers, mainly on grounds of worldwide inflation. Inflation did not occur suddenly with the beginning of the October war; it had started long before. It appears that Iran, taking undue advantage of the war and the embargo, and in association with other leading oil producers of the region, spearheaded the move to quadruple oil prices. As one researcher in the petroleum industry notes, "In the absence of the October War and the resulting Arab oil embargo, prices would not have risen so rapidly in so short a period."[37]

The price of oil is dubious and confusing. Behind this is the system of oil pricing which has developed over the decades. At the base is the posted price, which is a point of reference for tax and royalty collection and the buy-back price. The royalty is a fixed percentage of the posted price, which the oil companies pay to the host countries for extraction of oil. After the royalty has been paid and the cost of production, which now is about 25 cents per barrel (it was only about 11 cents in 1973), is deducted from the posted price, taxes are taken. The buy-back price was introduced after the oil embargo when some of the producing countries acquired 60 percent of oil company ownership. The buy-back price is set at a portion of the posted price.[38]

Often, the oil-producing countries decide to increase taxes, royalties, and buy-back prices, claiming that they are not raising prices. Companies making exorbitant profits from oil sales are not willing to reduce their profits. Thus, the increase is passed on to consumers.

Some readjustments have been made by Saudi Arabia, the UAE, and Qatar. They reduced their posted prices by 40 cents per barrel but raised the tax, royalty, and buy-back prices to almost 40 cents. As a result, there was no impact on the oil market and consumers continue to pay higher prices just as before.

By the time the first year of the energy crisis neared an end, the Arabs launched a vigorous worldwide campaign to explain their oil policy. They clarified their stand on the use of oil in their struggle for the liberation of Israeli-occupied Arab territories.

Not content with this, in full-page advertisements in the *Washington Post* and the *New York Times* on the eve of the new year, the Saudi minister of state for foreign affairs wished the U.S. a happy new year, on behalf of the Arab people. Appreciating the fact that the United States' "holiday season may have been marred by the hardships of the energy crisis,"[39] the Saudi minister brought home to the United States the plight of the Arabs by saying, "ours is haunted by the threat of death and continued aggression."[40]

Referring directly to the Arab oil embargo the Saudi minister said:

> We cut oil supplies to the United States after the United States, which had repeatedly assured us of our rights to our lands, made massive arms deliveries to the Israelis to help them remain in our lands. We did so not to impose a change in U.S. policy in the Middle East but to demand the implementation of U.S. policy in the Middle East, as it has been repeatedly defined. We did so not to "blackmail" the American people, but to put our case to them as effectively as we knew how.[41]

Almost simultaneously, King Faisal, in his first public speech since the October war, called on all Muslims to mobilize their resources "to rescue our sacred places in Jerusalem from the Zionist and Communist menaces."[42] Speaking to a group of high-ranking foreign pilgrims, Faisal, who was the official protector of the holy places of Islam, told them that he had a special responsibility for liberating Jerusalem, including the mosque of Omar, Islam's third holiest—Mecca and Medina being the first and the second, respectively.

Worldwide Arab psychological warfare, justifying their use of oil as a weapon, only proved how vulnerable the Arabs themselves really were. The use of oil as a weapon created an anti-Arab backlash, which generated much public sympathy for Israel in Europe. Government policy in Europe called for refraining from any actions that might antagonize the Arabs, thus endangering the oil supply to Europe. Considering the neutral attitude of Western Europe and Japan, the Arab oil states decided to end their monthly production cuts and ease restrictions on oil exports to those nations.[43] In an editorial the *New York Times* said:

> The Arabs have brought their oil weapon into play in the Mideast war, but cautiously and in a manner that leaves the way open for diplomatic cooperation ... it is evident from the conduct of the Arab diplomatic delegation ... that the leading oil states are more interested in accommodation than in confrontation with the West....[44]

Saudi Arabia, taking the lead in imposing the oil embargo, began to plead for moderation soon thereafter. It was under the moderating influence of its oil minister, Yamani, that the embargo was lifted on March 18, 1974, by a decision of most of the OAPEC states meeting in Vienna. He also "steered OPEC towards a freezing of crude oil posted prices for another three months . . ."[45] Following the lifting of the embargo, oil supplies to the United States were resumed. Yamani accomplished a three-month price freeze by threatening to break the united oil price front of OPEC by posting a separate lower price. To maintain the unity of OPEC—a most successful international cartel—the members agreed to Yamani's proposal.[46]

There were two OPEC meetings in the summer of 1974: one in Quito, Ecuador, in June; the other was in Vienna, Austria, in September. In these meetings Yamani tried to stave off price increases by threatening to increase production. The other producers countered by threatening to cut production by an equal amount. Saudi Arabia felt that the price of oil was already high and any further increase would attract substitution of oil, making its reserves useless. Other producers disagreed with Saudi Arabia. The result was a compromise, a slight price increase.

The notion that Saudi Arabia, as the largest exporter of oil, could unilaterally bring down the price of oil is somewhat misleading:

An increase in export capacity by 2.5-3.5 million b/d which is all the country would be physically capable of in the next 3 years could be largely offset by corresponding reductions in countries with surplus oil revenues.[47]

Nevertheless, temporarily, the oil embargo proved to be the strongest of diplomatic weapons. "It prodded the U.S. which in turn prodded Israel, and the result was the disengagement of Israeli and Egyptian forces along the Suez Canal."[48] Although the Soviet Union, the principal arms supplier to the Arabs and their political booster, called on Arab oil producers to maintain the embargo, the Egyptians and the Saudi Arabians were eager to reward Henry Kissinger for his step-by-step effort to bring about a ceasefire, and subsequent withdrawal of Israeli troops from the west bank of the Suez Canal.[49]

Most people were puzzled by the Arab decision to use oil as a weapon, an act which seemed unfriendly to everyone. Apparently oil was used as a political weapon and not so much for economic expedience, although some OAPEC and OPEC nations took economic advantage of it initially.

THE ARAB STRATEGY

Soon after the embargo was lifted, the Arabs expounded at the United Nations General Assembly special session in April 1974 on their oil policy objectives. Algeria's president, Boumediene, who convened the session, developed a "work program in five points," which was the Algerian view of oil's role in the economic development of oil-producing countries. The five-point program, which included the nationalization of oil and other resources in the developed countries, was announced by Boumediene after wheat and fertilizer prices, which had doubled between June 1972 and September 1973, were reconsidered. Reacting to Western criticism of the oil price hike, Boumediene rebutted:

> ... the fundamental difference that explains the greatly dissimilar reactions caused by rises in fertilizer and wheat prices on the one hand, and in the price of oil, on the other, resides in the fact that the proceeds of the increase went to developed countries in the first case and to developing countries in the second ... [50]

Addressing the special session of the General Assembly, Saudi Oil Minister Yamani declared that oil-exporting countries would reject any attempt to impose a control on oil prices. He added that Saudi Arabia could afford to cut its oil production by half to raise the price further, but did not do so because of its "sense of responsibility towards the rest of the world."[51] In exchange for this goodwill he wanted the industrialized world to help set up diversified industrial structures in Saudi Arabia to assure income when the oil runs out.

The Arab strategy, since the imposition of the embargo, has been to create a breach in the relationship between America's NATO partners. While the United States was firm in its Mideast policy, Europeans were not in a position to offer effective resistance to Arab pressure, simply because of their greater dependence on Arab oil. Arab pressure on Europe seemed like a reversal of the earlier European colonial situation. Although no permanent dent was created in the United States' relations with its NATO partners, the tendency in Europe "was not to blame the Arabs, or even Israel, but to say that it was United States' Middle Eastern policy that was causing Europe to freeze this winter."[52] Indirectly, Europeans were putting pressure on the United States to adopt an evenhanded policy in the Middle East. It is through continued talks and closer relations with Western Europe that the Arabs attempt to weaken Israeli ties with Europe and the United States. If the Arabs succeed in this strategy, they will win and Israel will lose a contest which otherwise has not been resolved by war.

Another Arab strategy has been to isolate Israel from the main current of world politics and create a sense of insecurity within the Jewish state. The need for peace is greater for Israel than for the Arabs, both economically and politically, mainly because of three new developments after the 1973 war. Because of insecurity, immigration to Israel is slowing. Emigration is accelerating. Added to this is the inflation of Israeli currency because of greater military spending, resulting in the devaluation of the Israeli pound by 43 percent in 1974 and again by 2 percent in 1975. Monthly 2 percent devaluations were to take effect if the economy continued to decline. At the same time, foreign investment in Israel slumped by about 60 percent.[53] Fears of an Arab boycott inhibit foreign investment in Israel. Politically, the Arabs were effective in isolating Israel. Following the October war, many African and Asian nations, which had earlier established diplomatic links with Israel, severed those relations. Thus, Israel is becoming insecure, while the Arabs have tremendous newfound wealth in the form of petrodollars. As the Arab-Palestinians gain more sympathy in the world there has been a reduction of sympathy for Israel. The Arabs were successful in cutting off support for Israel by the United Nations Educational, Scientific, and Cultural Organization in 1974. In the future Arabs might initiate other such actions through the United Nations and its agencies.

> It can be said that Israel is in many respects alone in the world because its own orientation is against the general trend. Even American-Israeli relations are in the process of change, owing largely to the fact that the United States can no longer centre its Middle Eastern policies on the interests of Zionism alone.[54]

If Israeli isolation continues and the Arabs continue to gain power, prestige, and credibility, Israel will lose its position in the world.

The energy crisis achieved tremendous political and economic gains for the oil producers. Arab military success in the October war and their use of oil as a weapon showed that they were acquiring not only the tactics of modern warfare but also the art of diplomacy. Moderation has been constantly threatened by those in the Mideast who feel from experience that militancy pays.[55] It was for this reason Saudi Arabia set aside its policy of moderation and used the oil weapon. This move satisfied the nationalist aspirations of the Arabs. At the same time, Saudi Arabia and other nations realized that their weapon was a two-edged sword. "Oil is a double-edged weapon," said Habib Bourguiba, president of Tunisia, an oilless Arab nation in North Africa.[56] Before the weapon could cause irreparable damage to the U.S. economy, thus disturbing Arab economic interests, the embargo was eased. As an Arab minister stated:

> If the embargo were to remain, we would see a major recession in America. That, in turn, would affect all of us adversely. Our economies, our regimes, our very survival depend on a healthy U.S. economy.[57]

The early lifting of the embargo indicated that the Arabs, particularly the Saudis, decided not to abuse their economic power. "There is a growing realization that this power—based on hydrocarbons—is finite and that alternative sources of revenue should be developed speedily."[58] According to present indications, Saudi Arabia's oil reserves will dry up in the year 2039, making the desert kingdom an oilless country.[59]

Saudi Arabia's oil resources were developed exclusively by the United States. It is obvious that the Saudis need U.S. technology to set up a functional economic infrastructure on which a future oilless economy can be based. This is a gigantic task and, judging from the world's available technology, can only be delivered to the Saudis by the United States, Europe, and Japan.

Speaking from an economic viewpoint, the oil embargo has paradoxically strengthened the U.S. position in the Middle East, especially in Saudi Arabia. Saudi relations with the U.S. oil companies have undergone many changes since beginning operations in the 1930s, but although the principal target of the oil embargo was the United States, U.S. oil companies continue to operate in Saudi Arabia.

The energy crisis and the world's increasing dependence on Middle East oil have made oil a controversial subject and have created a climate of doubt and suspicion regarding the motives of the host government, its relationship with the oil companies, and company relations with their own countries.

A PALESTINIAN-ISRAELI ISSUE

The oil embargo was not intended to end the October war, but to achieve the evacuation of Israeli forces from all Arab lands occupied during the 1967 war. This indicated that the Arabs were willing to accommodate Israel within the original boundary given it by the United Nations in 1948. Another political reason the oil embargo was imposed was to restore the legitimate rights of the Palestinian people. Although it was not spelled out during the October war, the restoration of rights meant their return to an independent Palestine. This implied an end to Israeli occupation. As subsequent events showed, this was what the Arabs actually meant by the restoration of the legitimate rights of the Palestinians.

The Arab-Israeli conflict is essentially a Palestinian-Israeli issue. Palestinians who were driven out of their homeland in 1948 live as second-class refugees scattered all over the Arab world. Although they are Arabs, many Palestinians prefer to be known as Palestinians due to their nostalgic feeling toward their homeland.

The Palestinian movement was dramatized at the Rabat summit of Arabs in October 1974, where the Palestine Liberation Organization (PLO) won a major diplomatic victory when 20 Arab nations backed its claim as the sole representative of the Palestinian people, despite King Hussein's opposition. He, however, succumbed to the unanimous pressure of the Arab world.

The United Nations, which, in effect, drove the Palestinians out of their homeland through a 1947 resolution creating Israel, bent its tradition by inviting the head of the PLO to address the General Assembly as though he represented a formal government.[60]

Yasir Arafat, the head of the PLO, while addressing the United Nations on November 13, 1974, took an all or nothing approach that culminated in the passage of a resolution on November 22 recognizing the rights of the Palestinians to independence and sovereignty in Palestine.[61] The General Assembly supported the resolution by 89 to 8 with 37 abstentions, mostly from Europe and Latin America. The United States, Israel, Norway, Iceland, Bolivia, Chile, Costa Rica, and Nicaragua voted against the resolution.[62]

Of all the developments in the Middle East since the creation of Israel, the passage of the Palestine resolution is the most significant. This is because the United Nations, which created Israel by passing a similar resolution, passed a counterresolution to undo Israel and recreate Palestine. In a simplistic sense Palestine has achieved its international personality, being endorsed by the world in a landslide vote. Restoring Palestine and Palestinian rights is difficult to accomplish. As happened in the case of Israel, a United Nations resolution to partition Palestine into Jewish, Arab, and international sectors was implemented by successive wars.

PALESTINIAN THINKING

The Palestinians are known as people with a chip on their shoulders. Their image is that of a small "nation," half of which lives in Israel, the other half scattered around the world, locked in perpetual conflict with Israel, the strongest military power in the Middle East.

While this image evokes sympathy and admiration from those who understand the grievance the Palestinians are struggling against great odds to redress, it generates antipathy among those who lack sensitivity to their plight. The Palestinians tend to be perceived either as heroic freedom fighters bravely defying injustice or as irrational rebels working against the established order.

A brief survey of Palestinian thinking reveals three important desires: (1) liberation and return, (2) establishment of a democratic community, and (3) a Palestinian state.

Liberation and Return

The transformation of Palestine into Israel was not a simple or harmless political act. Nor was it a coup that transferred political power from the indigenous Arab majority to the Jewish minority. It was a traumatic and violent social upheaval that involved the destruction of Palestine and its society. It turned history inside out for the Palestinian people, dooming them to national dispossession and exile in an age of national resurgence and independence.

The Palestinian people, naturally and instinctively, reacted to this dual injustice of dispossession and exile by seeking redress through the liberation of their occupied homeland and the repatriation of their exiled community. Hence, during the decades that followed the 1948 destruction of Palestine, Palestinian political thought has been symbolized by the twin slogans, "liberation" and "return."

During the initial phase, the Palestinians had a moral position rather than a political program. They simply confronted the injustice they suffered with its obverse: liberation and repatriation.

The Democratic Community

By the mid-1960s, the Palestinians had managed to evade oblivion through an unparalleled educational drive that, within one generation, turned them into a highly educated and skilled community. Consequently, they became active in the cultural and economic revival of the Arab world. Furthermore, they managed to rehabilitate their shattered society by creating Palestinian institutions that preserved their identity, instilled pride, and provided them with needed services in many areas of life. This process of social regeneration was consummated in 1964 with the establishment of the Palestine Liberation Organization, which became the institutional expression of resurgent Palestinian statehood.

These developments made it inevitable that Palestinians would return to the struggle over their fate and the future of their country. They began to take the struggle more seriously, and their national liberation movement was given more importance. The Palestinians felt a need to develop their grievances into a cause, and turn their moral position into a political program.

The most troublesome problem they faced was how to reconcile their national rights with the demographic and political realities of the destruction of Palestine and subsequent events. The Palestinians formed a creative response. They advocated the reconstitution of Palestine into a democratic, nonsectarian republic. According to this concept, usually labeled the "democratic secular state," Palestine would be reunited to become the common homeland of its indigenous Palestinian-Arab people as well as the emerging Israeli-Jewish community, where citizenship and sovereignty would be shared without religious or ethnic discrimination. This proposal became official Palestinian policy and was formally endorsed by the PLO's lawmaking body, the Palestine National Council, in its 1969 meeting and in subsequent sessions.

The Palestine State

The concept of a pluralist society clashed with Israel's Zionist ideology. Israel rejected the democratic nonsectarian state idea immediately and categorically, dubbing it a euphemism for the destruction of Israel, and made it illegal for its citizens to support or advocate the idea. In 1974, the PLO was forced to relegate the democratic nonsectarian state idea to its inactive files, calling it a long-range solution to the Arab-Israeli conflict.

Meanwhile, a broadly based international consensus emerged, calling for the creation of a Palestinian state in a fraction of Palestine as the basis for resolving the Arab-Israeli conflict. The PLO, in a second historic political shift, supported this international consensus. Beginning in 1974, the Palestine National Council adopted a series of resolutions putting the PLO on record as favoring a political settlement based on this "two-state solution" to the conflict. It endorsed United Nations resolutions to that effect and supported a similar formula for Arab-Israeli peace, which was approved unanimously by the Arab states at their summit conference at Fez, Morocco, in 1982.

These successive shifts in Palestinian political thinking reflect concern for the plight of the Palestinian people and the well-being of the people of the Middle East. These shifts do not signify erosion of Palestinian conviction in the justness of their cause or a wavering of their belief in the legitimacy of their struggle. Moreover, they are not based on weakness. On the contrary, the most compelling evidence that Palestinian flexibility was motivated by a desire for peace is that these policy shifts occurred in moments of strength.

The democratic nonsectarian state proposal was made in 1969, during the euphoric period of Palestinian resistance. The call for a Palestinian state came in the aftermath of the 1973 war when, for the first time, the Arabs fought Israel to a standstill.

THE AFTERMATH OF THE OIL WEAPON

The aftermath of the use of the oil weapon strengthened the Palestinian case. It received an unexpected boost when the late shah of Iran, who had shown new support for the Arabs, joined Sadat in calling for Israel's total withdrawal from occupied Arab land by affirming the rights of the Palestinian people in their home-land.[63]

As time passed, the effect of Arab petrodiplomacy was felt more in the international economic field than in the political arena. This happened as a result of oil price escalation, for which the Arabs share little blame. Although Saudi Arabia led the Arab nations in imposing the embargo, there probably was no intention of increasing oil prices suddenly. Nevertheless, the situation was soon out of control and, taking advantage of the embargo, Iran and Venezuela took the lead in escalating the price. The quadrupling of oil prices in little more than a year is perceived to be the cause of today's economic chaos, but this is not the whole cause. The sudden fourfold increase by OPEC and the back-door method by which it was made have caused the greatest disarray in the world today.

An interesting phenomenon is the radical change in the world oil market. Since the last embargo, the non-Arab non-OPEC flow of oil in international trade has increased. It is possible that this oil will minimize the shortage and vitiate the political impact of any future embargo.

The Arabs may not be able to hamper oil to embargoed countries because oil is handled by a variety of operators. It was indeed difficult for the Arabs to enforce the destination embargo in 1973. Furthermore, consumer adjustment is much easier now, as it will be in the future. The world will not be taken by surprise.

The oil crisis has brought the separate pieces of oil politics to-gether in a dramatic global collage and calls for measures that would assume all oil consumers of acceptable. terms and conditions. We have been watching an erosion of the world's oil supply and financial systems comparable to the depression of the 1930s.

Whatever may be the overall effect of these developments, the threat of another embargo will exist until an Arab-Israeli settlement is reached. The use of oil as a diplomatic weapon is almost certain in the event of war. Even without a war, a breakdown in negotiation or Israeli retaliation against Arab guerrillas in Lebanon and elsewhere could trigger an oil embargo. With about one-third of world oil ex-ports currently coming from the Arab world, it is obvious that all oil-importing countries have a vital interest in the improvement of Arab-Israeli relations.

The growing disparity between oil haves and have-nots has been conducive to international financial chaos and misunderstanding. This inequality has been accentuated because international oil is now owned by a few developing nations which are still poor compared to Western nations. Naturally, they jealously guard their newfound wealth against outside encroachment. Not all jealousies and conflicts are favorable for a quick solution. Even though the oil-producing and -exporting nations do not see it this way, no return to the old order is possible. The transfer of financial power and its components has already taken place and the world has to adjust to the new reality, as the Arabs call it. The only danger is in extremism. Clearly an oil embargo and the subsequent explosion in oil prices constitute an extreme action.

NOTES

1. Sydney N. Fisher, *The Middle East* (New York: Alfred A. Knopf, 1959), p. 583.

2. *Newsweek*, April 9, 1973.

3. Ibid., December 30, 1974.

4. Richard H. Ullman, "After Rabat: Middle East Risks and American Roles," *Foreign Affairs* 53, no. 2 (January 1975): 287–88.

5. Ibid., pp. 289–90.

6. International Institute for Strategic Studies, *Strategic Survey, 1973* (London, 1974), p. 1.

7. George Lenczowski, *Oil and State in the Middle East* (Ithaca, New York: Cornell University Press, 1960), p. 188.

8. Ibid.

9. Ibid.

10. Leonard Mosley, *Power Play: Oil in the Middle East* (Baltimore: Penguin Books, 1974), p. 343.

11. Ibid.

12. Ibid., p. 344.

13. *Middle East Economic Survey* (Beirut), August 11, 1972.

14. Ibid., July 21, 1967.

15. Ibid.

16. *U.S. News and World Report*, January 15, 1973.

17. Malcolm H. Kerr, *The Arab Cold War* (London: Oxford University Press, 1971), p. 129.

18. Nadav Safran, "The War and the Future of the Arab-Israeli Conflict," *Foreign Affairs* 52, no. 2 (January 1974): 219.

19. See an English review by Burhan Dajani of the original study in Arabic entitled "Economic Interests in the Service of Arab Causes" in *Journal of Palestine Studies* 3, no. 1 (Autumn 1973): 142–44.

20. *Washington Post*, June 17, 1973.

21. *Christian Science Monitor*, September 4, 1973.

22. Arab Oil Ministers' Resolution, October 17, 1973 (Washington, D.C.: Embassy of Kuwait).

23. Ibid.

24. Safran, "The War and the Future," p. 220.

25. Ibid., p. 221.

26. Dana Adams Schmidt, *Armageddon in the Middle East* (New York: John Day, 1974), p. 212.

27. Ibid.

28. *Economist* (London), November 3, 1973.

29. *Daily Telegraph* (London), November 17, 1973.

30. *Sunday Times* (London), November 14, 1973.

31. Schmidt, *Armageddon*, p. 213.

32. John H. Lichtblau, "Arab Oil and a Settlement of the Middle East Conflict," in *After the Settlement: New Direction and New Relationships* (a paper delivered at the 28th Annual Conference of the Middle East Institute, Washington, D.C., October 12, 1974).

33. Ibid.

34. Ibid.

35. *Washington Post*, December 24, 1973.

36. Lichtblau, "Arab Oil."

37. Ibid.

38. Adapted from Joseph Kraft, "The Rising Price of Oil," *Washington Post*, December 1, 1974.

39. *Washington Post*, December 31, 1973, and *New York Times*, December 31, 1973.

40. Ibid.

41. Ibid.

42. *Washington Post*, December 31, 1973.

43. Ibid., December 26, 1973.

44. *New York Times*, October 18, 1973.

45. *Middle East Economic Digest* (London), March 22, 1974.

46. *Washington Post*, March 28, 1974.

47. Lichtblau, "Arab Oil."

48. *Newsweek*, March 25, 1974.

49. Ibid.

50. *Arab Oil and Gas Journal* (Beirut), May 1, 1974.

51. Ibid.

52. Schmidt, *Armageddon*, p. 216.

53. *Washington Post*, April 2, 1975.

54. Alan R. Taylor, "The Isolation of Israel," *Journal of Palestine Studies* (Autumn 1974): 93.

55. *Middle East Economic Digest* (London), March 22, 1974.

56. Mosley, *Power Play*, p. 426.

57. *Newsweek*, March 25, 1974.

58. *Middle East Economic Digest* (London), December 28, 1973.

59. Mahdi al-Bazzaz, "Middle East Oil Revenues: An Assessment of Their Size and Uses," *Middle East Economic Digest* (London), March 15, 1974.

60. *New York Times*, November 16, 1974.

61. *Washington Post*, November 23, 1974.

62. *United Nations General Assembly Provisional Verbatim Record* No. A/PV. 2296 (New York, November 22, 1974), p. 46.

63. *Washington Post*, January 12, 1975.

7

Endless Trouble and Turmoil in Lebanon

Lebanon consists of irreconcilable religious and ethnic minorities that live under varying degrees of hatred and hostility toward one another. The current Lebanese crisis is attributable to the heterodox character of the country's four million people, who embrace seven major and at least ten minor sects, each of which demands social and cultural distinction.

The estimated population of Lebanon in 1985 is:[1]

Group	Estimated Population
Shiite Muslims	1,000,000
Maronite Christians	600,000
Sunni Muslims	600,000
Palestinians	500,000
Greek Orthodox	400,000
Druze	300,000
Melchites	250,000
Armenians	250,000
Protestants (and other minorities)	100,000

Lebanon's Arab character and the Arabic language are two factors holding the country together. Divisive pulls, though often latent in the past, have a logic of their own. A sizable Palestinian presence in the country and frequent Israeli military forays and invasions have played a catalytic role in giving internal differences a sharp edge. These differences culminated in the Lebanese civil war, which began on April 13, 1975. For the past decade Lebanon (especially Beirut) has been torn apart. Particularly, "The 1982 Israeli invasion of Lebanon showed how wide the Gulf between nation and state in the Arab world has grown."[2] The Israelis came with a bang and went home in June 1985 with a whimper, in the end slipping quietly back across the border, leaving a Lebanon bloodied by unending turmoil.

HISTORICAL DEVELOPMENT

The strength the divisive forces have shown has its roots in historical developments dating back to 1096, when the occupation of the Christian holy places in Palestine by Caliph al-Hakem led to a series of eight campaigns, known as the Crusades (1096–1291). These campaigns were undertaken by the Christians of Europe to recover the holy land from the Muslims. The Crusade was proclaimed by Pope Urban II at the Council of Clermont-Ferrand in France. After taking Jerusalem, the crusaders turned their attention toward the Lebanese coast. Tripoli capitulated in 1109; Beirut and Sidon, in 1110. Tyre resisted stubbornly but capitulated in 1124 after a long siege.

Although the crusaders failed in their objective, they left important impressions in Lebanon. Conspicuous among them are the remains of many towers along the coast, ruins of castles on hills and mountain slopes, and numerous churches. Of all the contacts established by the crusaders with the peoples of the Middle East, those with the Maronites of Lebanon were among the most effective and enduring. The Christians acquainted the Maronites with foreign ideas and made them more receptive to friendly approaches by Westerners. During this period the Maronites were brought into a union with the Holy See, which still survives. France was a major participant in the Crusades, and French interest in the region and its Christian population dates from this period.

Bitter conflicts among regional and racial groups in Lebanon and Syria characterized the thirteenth century. The crusaders, Mongols from the steppes of central Asia, and Mamluks from Egypt all sought to control the area. In this hard and confused struggle for supremacy, victory came to the Mamluks.

The Mamluks were Turkoman slaves from the Turkmenian area east of the Caspian Sea and Circassian slaves from the Caucasus Mountains between the Black Sea and the Caspian Sea. They were imported by the Muslim Ayyubite sultans of Egypt to serve as bodyguards. One of these slaves, Muez-Aibak, assassinated the last Ayyubite sultan, al-Ashraf Moussa, in 1252 and founded the Mamluk sultanate which ruled Egypt and Syria for more than two centuries.

During the eleventh through the thirteenth centuries, the Shiites, an Islamic sect, settled in the northern part of Bekaa Valley and in Kasrawan, the mountains northeast of Beirut. They and the Druze rebelled in 1292 while the Mamluks were busy fighting European crusaders and Mongols. After defeating them the Mamluks crushed the rebellion in 1308. To escape Mamluk reprisals and massacres, the Shiites abandoned Kasrawan. The Maronites, seizing the opportunity, moved from the Qadisha Valley and settled in Kasrawan. Mamluk rule was ended by the Turks.

The Ottoman Turks were a central Asian people. They served as slaves and warriors under the Abbasids. Because of their courage and discipline, they became the masters of the palace in Baghdad.

The principal European powers of the time, Britain, Austria, Prussia, and Russia, opposing the French pro-Egyptian policy, signed the London Treaty with Turkey on July 15, 1840. In this treaty they agreed to call upon Mohammad Ali, the founder of modern Egypt, to evacuate Syria. Upon Ali's refusal, Ottoman-British troops landed on the Lebanese coast on September 10, 1840. Confronted with such strong opposition, Mohammad Ali called off his campaigns. On October 14, 1840, Bashir II surrendered to the British and was exiled.

The Ottoman sultan proclaimed Bashir III as prince of the country, then known as Mount Lebanon. Under the new prince, bitter conflicts developed between Christians and the Druze. The sultan deposed Bashir III and appointed Omar Pasha governor of Lebanon. This appointment aggravated existing dissensions. In Turkey, European representatives proposed that Lebanon be partitioned between Christians and the Druze. The sultan adopted the proposal on December 7, 1842, and charged Assad Pasha, governor of Beirut, with dividing the country into a northern district under a Christian subgovernor and a southern district under a Druze, both officials to be responsible to the governor of Sidon, who lived in Beirut. The Beirut-Damascus road divided the two districts.

This partition proved to be unfortunate. Animosities, nurtured by French support of the Christians, by British support of the Druze, and by Ottomans seeking to abolish the semiautonomous status of Lebanon, resumed at a quickened pace.

Foreign interests in Lebanon transformed these political and social struggles into bitter religious conflicts, leading to massacres of the Maronites by the Druze in 1860. These events offered France an opportunity to restore order and tranquility. To thwart French intervention, the Turkish govenment sent Fuad Pasha, minister of foreign affairs, to Beirut with plenipotentiary powers. In September 1860 he informed the French general, de Beaufort, that severe measures had been taken against the instigators of the recent events in Lebanon.

An international commission was formed on October 5, 1860, by five European powers—France, Great Britain, Russia, Austria, and Prussia—and Turkey. This commission had two objectives: to investigate the causes of the 1860 events and to propose a new administrative and judiciary system for Lebanon to prevent the repetition of such events.

The commission members agreed that the 1842 division of Lebanon between the Druze and Christians had caused the religious disturbances of 1860. They reunited Lebanon under a Christian governor appointed by the Turkish sultan. The governor was to be assisted by an administrative council composed of representatives of different religious communities in Lebanon. The new system brought some stability to the Lebanese province of the Ottoman Empire.[3]

The Ottoman Empire's tenuous hold over Lebanon came to an end with its collapse at the end of World War I. Lebanon was passed quickly into French hands by the League of Nations. In 1920, the French decreed the restoration of Lebanon to its previous boundaries. While some Muslim leaders would have preferred to see Lebanon incorporated into Syria, the Christians had no negative disposition toward the French policy.

CONTEMPORARY PERIOD

In the context of the Lebanese political formula, power is granted to the executive branch according to the 1926 constitution.[4] As a result of national and international pressure, France granted Lebanon's independence on November 26, 1941.

Following independence Lebanon, through the unwritten National Covenant of 1943, established a unitary and parliamentary government which had a chamber of deputies. Until 1952, the number of Christian deputies (of various sects) was 44. Muslim deputies numbered 20. In 1960 the number of Christian members of parliament became 54 and Muslims were more than doubled to 45. This division of seats in the parliament is accompanied by an equal division of government and army posts. The president is traditionally a Maronite, the prime minister a Sunni, the speaker of the house a Shiite, the deputy speaker an Orthodox Christian, and so on.

The allocation of government and parliamentary seats clearly favored the Christians because a slight majority existed then. Now the Shiites are more numerous than the Maronites. Besides, a "Government thus constructed was designed to symbolize fair allocation rather than to act, and was often immobilized in its respect for sectarian balances and susceptibilities. . . ."[5]

The population forms a mosaic of religious communities. There are seven major religious groups—four Christian, two Muslim, and one Druze. In addition, there are ten smaller groups. Nearly all Lebanese belong to one. The last official census was in 1932, according to which Christians made up about 53 percent of the population; Muslims, 39 percent; and adherents of other faiths, the rest. Although the Muslims have now surpassed Christians numerically, for the sake of internal political stability, the government assumes these ratios still hold true. Religion and politics are intimately associated. Seats in the chamber of deputies and cabinet are distributed on the basis of proportional representation among the major religious groups. This policy is known as confessionalism.

Lebanon was not long ago being cited as a model of intersectarian balances. When the Palestinians talked of a united secular and democratic Palestine with Muslims, Jews, and Christians living happily together, they presented Lebanon as an example of such cooperative coexistence. Today Lebanon is on the verge of disintegration. At the core of this crisis lies the demand for the abrogation of the outdated covenant.

But the Christians and the Muslims conceive the Lebanese state in different ways. Most Christians do not want to be embroiled in Arab affairs and lean toward the Western world. Most Muslims consider Lebanon to be part of the Muslim world and favor active involvement in its politics.

Along with this dual political orientation runs economic opposition. The Maronites are the richest segment of the population and have gained the most by Lebanon's banking and trade boom—a result of the country's semiisolation from Arab affairs. The Maronites are identified with the right wing, which favors detachment from Arab politics, and who tend to disrupt Lebanon's commercial and financial status. The Muslims, particularly the Shiites, are have-nots, with leftist inclinations. Political domination by the Maronites is said to militate against the economic interests of Muslims and the Arab identity of Lebanon.

The difficulty in maintaining an equilibrium between the two main sections of Lebanon's population has been aggravated by Lebanon's peculiar position in inter-Arab affairs, especially in moments of crisis in the Arab world.

Under Nasser the issue of Lebanon's position came to the fore in the context of increasing radicalization of the Arab world.[6] In the wake of the 1956 Suez crisis, President Camille Chamoun of Lebanon, with solid support from Lebanese Christians, opposed Lebanon's total commitment to the Cairo-led progressive Arab bloc. In May 1956, violence flared in an upsurge of Muslim discontent. The event coincided with a left-wing revolt in Baghdad, leading to the violent and bloody overthrow of the pro-Western Hashemite government. Chamoun, afraid of a similar fate for his government, appealed to the United States for assistance and protection. This prompted President Eisenhower to send U.S. troops to Lebanon in 1958. In June 1967 during the six-day Arab-Israeli war, Lebanon again adroitly managed to avoid committing itself to a militant posture.

THE PLO ELEMENT

Against this background of the two-sided Lebanese political system and of the country's peculiar position in the Arab world, it is easy to understand why trouble erupted when Palestinian guerrillas introduced a new element of tension into the Lebanese situation.

When the PLO's quarrel with King Hussein of Jordan ended in 1971, through the mediation of countries like Egypt and the Sudan, the Palestine resistance movement faced a grave problem. Its very existence was threatened. The movement had lost its bases in Jordan and other Arab countries.

Lebanon's position on the PLO underwent substantial modification from 1967 to 1973. At the outset, the government of Lebanon sympathized with the guerrillas and allowed them to arm themselves and function freely. The Israelis responded with repeated warnings that Lebanon would be held accountable for guerrilla actions. Thus, mixed feelings of sympathy for the guerrilla position and concern about Israeli raids into Lebanon warped the Israel-Lebanon relationship.

The tensions resulting from the guerrilla presence in Lebanon were manifest in the events of late December 1968. On December 26, two Palestinian guerrillas of the Popular Front for the Liberation of Palestine (PFLP), who were also Lebanese nationals, departed from Beirut by air, proceeded to Athens, and there attacked an Israeli civilian airliner with bombs and machine guns.[7] The Athens raid resulted in the death of one passenger and the wounding of a stewardess, as well as causing damage to the aircraft. The PFLP issued a statement from Beirut claiming responsibility for the action. Israeli officials announced that Israel held Lebanon responsible since the guerrillas had come from Beirut and because the PFLP statement had been issued there.

Lebanon faced a dilemma. Should it permit guerrilla activity on its sourthern border and face the possibility of Israeli raids or suppress the guerrillas and possibly be faced with substantial internal dissent? Guerrilla activity against Israel continued. There were raids by Israeli commandos and continued warnings by the Israeli government that it held Lebanon responsible for permitting the guerrillas to use the border villages as bases for attacking Israel.

Israeli retaliatory raids did not halt guerrilla attacks against Israel. Skirmishes continued along the border throughout the summer and escalated to the point of precipitating another deep raid by Israel into Lebanon on September 5, 1970. Israeli armored forces searched the Arqub region for two days. This incursion led to a United Nations Security Council resolution that demanded immediate and complete Israeli withdrawal from Lebanon.

The Israeli incursions and the guerrilla losses in the war in Jordan in September 1970 curtailed guerrilla activity. Increased guerrilla infiltration and Israel's counterattacks continued in 1972. Following the Munich massacre of September 1972, in which 11 Israeli Olympic athletes were killed by guerrillas, Israel launched a series of raids into southern Lebanon. The United Nations Security Council sought to condemn the Israeli action without reference to the preceding terrorist acts. The United States was concerned that this was a one-sided approach and vetoed the resolution.

A raid on the home of the Israeli ambassador to Cyprus on April 9, 1973, and an attempted hijacking of an Israeli plane in Nicosia were followed by an Israeli raid into Beirut and on other guerrilla bases in Lebanon. An Israeli commando raid on Beirut on the morning of April 10, 1973, resulted in the death of three high-ranking guerrilla leaders.

Lebanon's position on the Arab-Israeli dispute was again highlighted when the October 1973 Arab-Israeli war erupted. While Egypt and Syria launched a major attack against Israel on the Suez Canal and in the Golan Heights, Lebanon's role was minimal. On October 9, 1973, just a few days after the start of the war, Israel bombed a radar station in Lebanon, near Beirut. The Israelis claimed that the sophisticated radar station was providing information to Syria, a charge the Lebanese denied. The PLO guerrillas used Lebanon's southern border as a position from which to shell Israeli settlements until Israel invaded Lebanon in 1982. Since then, PLO activity from Lebanon against Israel has almost ended. The guerrillas have been expelled from Lebanon.

LEBANON AND SYRIA

Unlike the Israelis, who invaded Lebanon, the Syrians were invited to come to Lebanon's aid early in the civil war. They came with the United States' understanding and over the objections of the Soviet Union.

Lebanon desired to maintain close relations with Syria and reap the benefits derived from their relationship. Lebanon sought to maintain the overland route for trade with eastern Arab states to ensure the flow of oil through Lebanon to the Mediterranean, and to continue Lebanese imports of food from Syria.

Despite these considerations, relations deteriorated and various drawbacks persisted throughout the ensuing period. Syria accused Lebanon of harboring exiles who plotted against the Syrian government. The duty rates on trucks entering Syria were increased. Although Lebanon protested that these levies were in violation of a Lebanese-Syrian economic agreement, they were not initially reduced.

After negotiations between the two countries in the fall of 1968, the economic agreement between them was revised and, among other matters, the trade levies were reduced. Lebanon pledged to limit the activities of Syrian political refugees.

In the spring of 1970, Syria warned Lebanon that an attempt to suppress the Palestinians would not be tolerated. After an Israeli incursion into Lebanon in May, Syrian troops moved into Lebanon but withdrew after recovering the wreckage and pilot of a Syrian plane that had crashed after an air battle over Mount Hermon. Lebanon's prime minister, Karami; the chief of staff; and Foreign Ministry officials went to Damascus for discussions with Syrian leaders, especially with Prime Minister Atassi. Syria continued to pressure Lebanon on the guerrilla issue throughout 1970, but toward the end of that year Syrian President Assad assured Lebanon that commandos from Syria would not interfere in Lebanese affairs. By the end of 1970 there was a rapprochement between the Syrian and Lebanese governments that increased trade between them and imposed tighter restrictions on guerrilla activity around Mount Hermon.

Syria closed its border with Lebanon on May 8, 1973, as a move to support guerrillas in Lebanon. The Syrians protested Lebanese air force attacks on guerrilla positions and clashes between the guerrillas and the Lebanese army. The closing of the border caused economic hardship for Lebanon. Once the border was closed, Syria insisted on a number of concessions before the border could be reopened. Syria sought agreement on labor laws and protection for the thousands of unskilled Syrian workers in Lebanon. The border was reopened on August 17, 1973, after the Syrian and Lebanese foreign ministers issued a communiqué in which it was noted that agreement had been reached on most of the issues in dispute.

The year 1976 was one of fateful decisions for Syria. The country became increasingly involved in Lebanon's civil war. Once Syrian troops entered Lebanon, the Lebanese government relied almost entirely on Damascus for major foreign and domestic political decisions. Even the U.S. policy in Lebanon from 1976 to the 1982 Israeli invasion was contingent on its support of Syria's role in Lebanon.

The Syrians were not aiming to destroy the PLO, rather to control it either by replacing Arafat as chairman or by sharply curbing his authority. Animosity between Arafat and Assad further complicated their relationship. In light of this it is not surprising that

Arafat decided to snub the Syrians by moving his base of operations to Tunisia rather than Damascus after his expulsion from Beirut.

A number of circumstances also helped Syria's principal objectives in Lebanon. Syria's goal is to dominate Lebanon in order to increase its influence in Middle East politics and make crucial decisions there. The June 1985 U.S. hostage crisis in Lebanon is the latest example of this strategy. It was only when Syria put pressure on the Shiite terrorists that they released the U.S. hostage via Damascus.

What are the consequences of Syria's prominence in the Middle East? Arab nations, particularly Saudi Arabia, resent Assad's attempt to scuttle the first Fez summit meeting in 1982. Many Arab nations are unhappy over Syria's support of Iran in its war against Iraq.

Assad allowed Arafat to escape to the northern Lebanese town of Tripoli where the majority of guerrillas were concentrated. But ultimately Assad crushed Arafat. The end of the PLO in Lebanon was forced by Assad, even though the Soviet Union and several Arab governments pleaded for compromise. Syrian control over the PLO strengthened Assad's hand, although no Lebanese group was at ease under Syrian domination.

THE ISRAELI INVASION

To secure its northern borders against PLO raids, to eliminate its military and political presence from Beirut, and to neutralize Lebanon through a peace treaty, Israel began its offensive on June 6, 1982, using the pretext of its London envoy's assassination. The Israeli war strategy was outlined in Israeli Defense Minister Ariel Sharon's statement made in the wake of the war before the Knesset (parliament). He said, "We are determined that [the PLO] will not continue to exist . . . if we stand firm in our demands. We shall create a triangle of peace consisting of Israel, Egypt and Lebanon—with open borders from Beirut via Jerusalem to Cairo."[8]

Israel had been preparing for a full-scale attack since the return of the Sinai to Egypt under the Camp David agreement. After having neutralized its strongest Arab foe, Egypt, by a peace treaty, Israel turned to Lebanon to strengthen the security of its northern borders. Lebanon was chosen for the new round for two reasons. First, being the major stronghold of freedom fighters and contiguous to Israel, it facilitated PLO raids on the border areas. Second, the presence of a large number of Christians in Lebanon—Israel's closest allies against the Syrians and the Palestinians—convinced Israel of the success of its mopping-up operations against the Palestinians in Beirut. Having adequately planned for the attack, Israel waited for the opportune time to strike. Egypt's isolation from the Arab world, Iraq's involvement in the gulf war, and Syria's growing alienation from the pro-U.S. Arab governments because of its support of Iran created the right moment to move.

It appears that Israel achieved its major goals. Politically, it obliterated the PLO from Beirut. Militarily, it destroyed the PLO's might by rooting out its base and dispersing its fighters. Strategically, Israel secured its northern borders and was steadily moving toward its ultimate goal of creating a peace zone by using its occupation as a bargaining chip for a peace agreement with Lebanon. But Israel was unable to conclude a peace treaty with Lebanon.

Israel, by winning the war, established its military superiority over the PLO and the Arabs. Its euphoria is apparent from Menachem Begin's statement: "There is no other country around us that is capable of attacking us . . . we have destroyed the best tanks and planes that the Syrians had . . . Jordan cannot attack us . . . and the peace treaty [with Egypt] stood the test of the fighting in Lebanon."[9] Having achieved its objectives, Israel would not agree to anything that might nullify its gains. Israel sought to consolidate them by laying down conditions for its pullout from Lebanon. Its demands in this respect included a security pact with the Lebanese government plus a "security zone" stretching 40 kilometers from its frontiers under the control of a joint Israel-Lebanon military commission; direct negotiations with the Lebanese government for normalization of relations; removal of all Palestinian guerrillas from the Bekka Valley and the Tripoli area; and the Syrian withdrawal.

The Israeli offensive has not wiped out the PLO but has weakened it. Israel wants the PLO to disappear from the political stage.[10] But as long as the Palestinians remain dispersed over the Arab world without a homeland, Israel will never be safe. The PLO is not merely a political or military organization; it symbolizes the national movement of deprived Arab people. Apart from giving a temporary relief to Israel, the evacuation of the PLO from Beirut has not ensured Israel's safety nor lessened Palestinian resolve to carry on its commando raids deep inside the occupied territories.

The immediate gains aside, the war proved costly for Israel in terms of men, money, and image at national and international levels. Having lasted for more than one week, the war, besides increasing the casualties, has also hit the economy.[11] This is evident from the commodity price and high inflation rate. The Israeli government spent $2.5 billion on the war effort.[12] Added to this were the recurring expenses of maintaining its military presence in Lebanon.

Although most Israelis supported the invasion, the Israeli government's role in the massacre of Palestinian refugees at Sabra and Shatilla shook the supporters' faith. While the people expressed their anger over the slaughter by holding a rally in Tel Aviv on September 25, 1982, members of the opposition and Begin's coalition resorted to overt criticism in the parliament. A series of abortive no-confidence moves against Begin included a demand for a judicial inquiry into the government's role in the massacre and the resignation of Israel's energy minister. So strong was the demand for a full judicial investigation that Begin had to yield.

There has been no threat to Prime Minister Peres since withdrawal of Israeli troops in June 1985. A majority of the people support his policy of securing the northern borders and keeping control over the West Bank and Gaza. Nevertheless, the loss of support may become dangerous if a wave of criticism gathers momentum.

The crisis dampened the support of Israel's friends. Several United Nations resolutions concerning the Lebanese crisis, adopted with increased majority and in two cases with unanimity, indicate Israel's growing isolation in the world. Particularly noteworthy is the impact on United States-Israeli relations, which appear to be waning. President Reagan's tough talks with Israeli leaders, his warning of grave consequences implying the threat of sanctions, his call for a freeze on new settlements in the West Bank and his support of United Nations Security Council resolutions testify to his displeasure.

FOREIGN AND MULTINATIONAL FORCES

Though basically confined to Israeli and Palestinian soldiers, the Lebanese war produced far-reaching regional and international repercussions. By enhancing Israel's sense of invincibility, by weakening the PLO, by shattering the myth of an Arab army, by humiliating the Arabs for their inaction, by undermining Lebanese unity, and by reducing the superpowers' credibility, the conflict equally affected the combatants, their supporters, and outside powers.

The presence of Israeli, Syrian, and multinational forces and revived internal clashes between rival factions spoke of a dark future for Lebanon's independent, unified status. Its existence as a unified, sovereign state now depends as much on the attitude of Israel and Syria as on the ability of the Lebanese government to forge unity among the diverse warring factions. What makes its religious diversity more explosive now is the resurgence of animosity and bitterness between the Druze and the Maronite Christians and between the rival Muslim groups. While Druze leader Waleed Jumblatt supports the PLO and opposes friendship with Israel, the Maronite Christians have remained Israel's strongest allies and have fought Palestinians with Israeli arms. The historical rift between the economically and politically predominant Christian minority and the Muslim majority has led to a decade of bloodshed, civil war, and the involvement of foreign forces. The war has once again unleashed the suppressed hatred between the rival groups as is manifest in the recurring clashes between the Druze and Christians, and between the rival Muslim factions in the Tripoli area. With his power base in the Phalangists, who were fully backed by Israel, President Bashir Gemayel found it difficult to contain their power. His failure to control the strong Phalangist militia compelled the Muslims to rebuild their own forces.

The internal tension in Lebanon provided Israel with a good excuse to prolong its stay. Employing a divide-and-rule strategy by guiding the rival communities into a collision course, Israel fully exploited the situation to its own advantage.

Eager to get foreign forces out of Lebanon, the weak and insecure Gemayel was unable to act swiftly and decisively. Besides the fear of adverse Muslim reaction, Gemayel was hampered in reaching an understanding with Israel lest the Arabs be displeased. While receiving Arab money needed to reconstruct and rebuild his country, Gemayel can hardly afford to alienate the Arab states. He has to retain Arab goodwill for the sake of the thousands of Lebanese working in the Arab world and contributing to Lebanese foreign exchange earnings. He also has to maintain good relations with Syria, which controls Lebanon's trade with the Arab hinterland. By closing its frontier and airspace, as it has in the past, Syria could create problems for Lebanon. Syrian and Israeli interests, growing animosity, and revived clashes between rival factions weaken chances of a speedy restoration of Lebanon's complete independence.

ARABS ABANDON THE PLO

The Israeli invasion set off a crisis in the Arab bloc. Their rulers were losing face in the Muslim world, their people were disillusioned over their passive role, their military inferiority was exposed, and above all, their stability and survival in the Middle East were in grave jeopardy.

After abandoning the PLO in the battlefield, the embarrassed Arab leaders tried to compensate by becoming hosts to PLO fighters, initiating new diplomatic moves and by putting fresh pressure on the West.[13] To atone further for their inaction in the war, 20 Arab countries came out with the Fez declaration demanding the creation of a Palestinian state in the territories lost in 1967. Without specifically mentioning Israel, the signatories thrust the responsibility of maintaining peace for all states in the region on the United Nations Security Council, thus implying conditional recognition of Israel. It appears that an attempt was made in the declaration to accommodate the extremist and pacifist viewpoints.

The Arab rightists, in their solidarity with the Palestinians, avoided antagonizing the United States. By calling upon the United States to recognize the PLO and by urging the Palestinians to accept Israel's right to exist, they tried to find a suitable course.

As a result of the Arab-Israeli-Palestinian conflict, Arab credibility has reached its lowest ebb. Syria, Iraq, and Libya, which have been vocal in their call for confrontation with Israel, could do little in the war to support the Palestinians.

Politically, Arab offensive capabilities have been greatly contained by the successful United States-Israeli diplomacy of isolating Egypt from the Arab world through the Camp David agreement. Due to its size and population, only Egypt could pose a formidable threat to Israel. Without this source of strength, other Arab states lack adequate force to confront Israel. Israel probably would not have invaded Lebanon had it not been convinced that the Arabs would not respond. By not helping the Palestinians in an hour of crisis, the Arab leaders have tarnished their public image. The masses have become disillusioned and rallies have taken place in several capitals, including

Cairo and Damascus. The sooner the Arab leaders show tangible political gains as a result of their policy, the better will be their chances for respect and survival.

THE CURRENT SITUATION

Throughout the late 1970s and early 1980s, Lebanon's Christian and Muslim militias gained in military and political strength. Today, the two most important groups challenging the authority of the state are the Lebanese Forces (LF) and the Druze/Amal (Hope) coalition. Both deserve scrutiny.

The ascendancy of the LF as a unified Christian military alternative to the central government in Beirut has dramatically altered the traditional political puzzle of Lebanon. While the LF does not represent all Christians, it does enjoy a monopoly in the Mount Lebanon area. Simultaneously, Muslims in Lebanon, particularly among the Druze and Sunni communities, fear and distrust the LF, given the latter's very close ties to Israel, especially after the Sabra and Shatilla massacres where Palestinians and Lebanese citizens were killed. Increasingly, the LF has become a force to be reckoned with, having considerable financial and military support from Israel. While initially the LF sought Israel's aid to rid Lebanon of the Palestinians and Syrians, and counted on the Jewish state to help them create a Christian state in Lebanon, they ended up with an occupation which will be in effect for the foreseeable future.[14]

Jonathan C. Randall reports how the relationship evolved between Bashir Gemayel and successive Israeli leaders, leading to Gemayel's election to the presidency of Lebanon.[15] It is important to add that the early 1980s regroupment of numerous Christian militias was not a diplomatic achievement, rather the result of a succession of bloody feuds which scarred, perhaps permanently, the Lebanese political environment. In their effort to unify all right-wing militias, Gemayel and his LF attacked Shamunist forces, killing as many as 500 fellow Christians, including many innocent bystanders.[16] This victory over Shamun came two years after the June 13, 1978 attack on Ehden and the assassination of Tony Franjiyah and his family.[17] These actions meant that all Maronites, at least, would have little choice but to obey Gemayel. By August 1980, the only Christian forces outside the LF were the Franjiyah Marada in northern Lebanon, under Syrian tutelage. In a recent article on the Lebanese Forces, Lewis W. Snider suggests that the LF's "policy orientation and some of the activities of the Front suggest a Lebanese-centered view of Lebanon's problems, politics and approaches to post-war recovery, that potentially speaks to a much broader constituency than just to the Christians."[18] As long as elements within the Muslim communities continue to fear the LF, it is safe to assume that the primary constituency of the LF will remain in the Christian enclave of Mount Lebanon. It is crucial to note that the Lebanese Forces' staunch pro-state identity cannot stop others from expressing similar nationalistic feelings. Prominent Muslim political figures are also capable and do express their patriotism. Therefore, the LF cannot enjoy a monopoly, not on patriotic grounds. While independent from the Kataib Party, the LF is the most dominant element. The Gemayel family does not dominate the LF and their reliability is questionable.[19] Undeniably, in 1985 this family was the leading political component of the Lebanese political and military scenes.

By collecting taxes and imposing conscription in its enclave, the LF has attempted to create an alternative to the legitimate Beirut government. The LF has seriously jeopardized the authority of the Lebanese army when it suited its purposes. Assuming that the Lebanese army and government eventually gain control over all of Lebanon, the fate of the Lebanese militias would be questionable. Will they demobilize peacefully and turn over their considerable military equipment to a potentially strong central government?

Undoubtedly, Christians and Muslims living in LF-controlled areas have enjoyed a modicum of public services not available in other parts of the country. Every attempt has been made to provide the population with needed services, though at very high prices. These social services could not be provided without strong military backing which could provide punitive action in the face of dissent. In addition, many of these goods and services reach Lebanon through illegal imports, thus further eroding the legitimacy of the central government.

Assuming that the Gemayel government could secure the withdrawal of all foreign forces from Lebanon, and succeed in restoring its authority over all parts of the country, it is difficult to predict whether the LF would relinquish its nonmilitary organization. With respect to the Lebanese armed forces, the LF could conceivably compete for recruits. It is safe to assume that the army would not tolerate this.[20] While a potential military confrontation between the LF and the Lebanese army is possible, the more serious aspect of this contest is the support base among Lebanon's Christians. In the absence of trust of the central government, would the Lebanese Christians ever turn back on the Lebanese Forces?

Extensive use of force against Lebanon's Druze and Shiite populations by the Lebanese army and its multinational backers created a severe polarization, leading Waleed Jumblatt and Nabih Berri to seek Syria's assistance. The Gemayel government placed itself in opposition to the Druze and Shiites following the signing of the ill-advised May 17, 1983 withdrawal accord with Israel. Without a consensus of all Lebanese factions, Gemayel played his U.S. card and lost. The accord may have been a U.S. foreign-policy victory, but it did little to assist Lebanon's political reconstruction. Opposition to the accord grew steadily among Druze and Shiite leaders, who received additional support from Syria. Syria was anxious to profit from the opportunity to voice an opinion on Lebanese affairs and remind Beirut that the road to Lebanon's peaceful reconstruction passes through Damascus.

Throughout the fall of 1983, the Lebanese army was pitted against Druze forces in the Shuf Mountains, and in and around Beirut against Shiite forces. Early victories led Shiite and Druze leaders to call on Muslim soldiers and officers to stop firing against their fellow citizens. In September 1983, 800 Druze soldiers deserted their command in Hammana and Bayt al-Din, including the chief of staff of the Lebanese armed forces.

By late January 1984, Beirut offered to reinstate and promote the Druze who had deserted the army.[21] While this action may have had a demoralizing effect on those who had remained loyal, Gemayel had little choice but to agree to the reinstatement because Beirut had already set a precedent by reinstating Major Saad Haddad, after having tried him in absentia and dismissed him from the army.

As the winter 1984 showdown progressed, Shiite and Druze militias took control of West Beirut with "40 percent of the army's 27,000 active fighting men having gone over to support the Muslim militias or who refused to take part in any further fighting against them."[22] These massive defections crippled the Lebanese armed forces. Washington's Lebanon policy also was rapidly changing. The resulting ambivalence, both in Lebanon and in the United States, increased Druze and Shiite confidence, leading them to demand major political changes, including Gemayel's resignation. Calls for U.S. pressure on the Gemayel government increased. Daniel Pipes, for example, has suggested that the United States use its military and economic aid as well as its diplomatic and moral support to extract from Beirut concrete steps toward reconciliation. Presumably such steps would include "taking a new census, opening government offices to leaders of the opposition forces, scrapping the six-to-five ratio of Christians to Muslims in Parliament, and holding new elections."[23] Once implemented, these steps might preempt a Syrian takeover in areas north of the Awali River and preserve the territorial integrity of Lebanon. Invariably, these suggestions include a hope for an ideal peaceful resolution of the Lebanese conflicts. Conversely, however, the logic behind these assumptions rests on the premise that all Lebanese are willing to work for a united Lebanon.

It is possible to assume that the Phalange party, controlling East Beirut and much of the Mount Lebanon region, would be satisfied with seceding from the republic, should that solution guarantee the security of the area. The result of this potential occurrence could undermine the president of the republic. Gemayel's preferred position of increasing the strength of the Lebanese armed forces and uniting the public behind an institution whose legitimacy may be acquired fairly rapidly may offer a salutary avenue. In order to accomplish this, however, both Syria and Israel must refrain from interfering in the reconstruction of the Lebanese armed forces.

The Christians, who have dominated Lebanon's political life since it gained independence from France, are now faction-ridden, isolated, and being squeezed by their Muslim countrymen.

IMPACT ON THE UNITED STATES

The Lebanese crisis has eroded the credibility of both superpowers. If U.S. prestige in the Arab world has suffered due to its failure to prevent an Israeli invasion, so has the Soviet image through the defeat of its ally. Neither power has gained much influence over the other. Whatever strategic gains the United States might have won because the Soviet Union failed the PLO and Syria have been neutralized by its lowered image in the Arab world. The impact on the United States could be assessed in terms of its relations with Israel and the Arab states, and its attitude toward the Palestine problem.

President Reagan's apparent toughness with Israeli premiers Begin and Peres and their shared perception of the PLO compels one to reason that Israel would not have launched a full-scale offensive on Lebanon without U.S. approval. In view of their common interest in eliminating the PLO from Beirut and eventually from Lebanon, the possibility of their collusion could not be ruled out. Some Arab leaders, including the PLO chief, have held the United States responsible for the whole episode. Besides being accused of encouraging the 1982 invasion, the United States was blamed openly by Arafat for the massacre at the Sabra and Shatilla camps. The United States' failure to protect Palestinian refugees has lowered its esteem among Arab pacifists who for years relied on it as the only force able to secure Israel's recognition of Palestinian rights. Doubts have now been publicly expressed in Riyadh and Cairo about U.S. credibility. That Arab perceptions of Washington are undergoing a change is seen in the moderates' dissatisfaction with Washington's approach. Their firm stand at the Fez summit speaks of their changed mood.

The Reagan administration is trying an evenhanded approach. It sought to make Begin realize that U.S. "support for Israeli security . . . does not mean" that he "has a blank check from the U.S."[24] To mollify the Arabs and prevent Moscow from exploiting their bitterness, Washington made certain moves, the most important being its new peace initiative. By calling for Israeli withdrawal from the West Bank and Gaza, and self-government by the Palestinians of these areas in association with Jordan, Reagan has begun an approach different from his predecessor's Camp David formula. That his plan is an improvement on Camp David is also apparent from the initial positive reaction from Riyadh, Amman, and Cairo. Even the PLO chief considered it. In a bid to give a fresh look to his Middle East policy, Reagan appointed George Shultz as the new secretary of state because of his balanced attitude on Arab-Israeli conflict.

United States efforts to disengage its Middle East policy from Israel would not go beyond a moderate level without undermining its interest. This is obvious from its opposition to Arab attempts to remove Israel from the United Nations, its veto of a Soviet-sponsored resolution calling for an arms embargo against Israel, and from its reluctance to resort to economic or military sanctions against the same. Imposition of sanctions is one of the most effective means of bringing about a change in Israel's attitude.

What determines the United States' obvious tilt toward Israel is Israel's strategic and political importance in U.S. global strategy against the Soviet Union. Israel is the only major U.S. stronghold from which to counter a Soviet thrust in the Middle East. Their convergence of interests and shared perceptions regarding the Soviet Union have been made formal by a strategic consensus agreement. Neither Egypt nor Saudi Arabia can be, for varying reasons, as reliable to the United States as Israel has been. What further enhances Israel's importance to the United States is its democratic structure. No less decisive is the substantial weight and pressure of affluent Jews over the U.S. Congress and the White House.

However, the United States has been gradually isolated from the Arab world and the entire Muslim bloc. The alienation of Arab states would be a diplomatic gain for Moscow, which might eventually attack U.S. interests in the area. To improve and retain its diminishing influence, it has to bring Israel to accept Palestinian statehood. Being a democratic state itself, with firm conviction of human rights and fundamental freedom, its denial of these rights to the Palestinians neither conforms to its democratic traditions nor supports its role as the leader of the free world. The measures the United States has taken so far are too inadequate to evoke a favorable response in the Arab world or produce apprehension in Israel. The exigencies of the time demand unequivocal recognition of the Palestinian right to an independent homeland. By taking a bold and clear stand on the issue, Washington could cause a policy review in Israel.

The United States is increasingly becoming aware of the inevitability of a Palestinian state. A special Department of State advisory group on Middle East policy noted that since "a Palestinian state is virtually inevitable," the most sensible policy would be for the United States and Israel to help form it.[25] The advisory group opposes an independent state of Palestine, presumably because of a belief that an anti-U.S. majority in the PLO's rank and file would not only pose a threat to Israel but might serve Soviet interests. To avoid this it wants to establish a semisovereign Palestinian state controlled by King Hussein's rightist regime.

MOSCOW'S LIMITED OPTIONS

To keep Moscow's influence in the area minimal, Washington wants to exclude it from the Middle East settlement. Moscow, on the other hand, seeks to have a say. To ensure this, the Soviet Union, apart from denouncing the United States' desire for separate deals with the disputing parties, has been reiterating the urgency of an international conference. Its denunciation of the Camp David agreement and its promotion of the Brezhnev formula put forward in the aftermath of the Lebanese war reflect its policy objectives.

Being the major Palestinian arms supplier and a supporter of an independent Palestinian state, Moscow has an edge over the United States in its relationship with the PLO. However, its less assertive role in the conflict has damaged its credibility. The defeat of the PLO and Syria is a matter of discomfort. For one, it has established the superiority of U.S. arms over the Soviet weapons used by the PLO and Syria. For another, it provides an opportunity for anti-Soviet elements to point out inaction to prove that Moscow abandons its allies in their hour of need.

With these limited options, Moscow could not go beyond condemning the Israeli onslaught, urging the United States to act jointly to secure Israeli compliance with the United Nations Security Council decision, and launching its own Brezhnev plan. Soviet passivism can be attributed to three things. First is the Soviet Union's desire to avoid direct confrontation with the United States. Its intervention on the PLO's or Syria's behalf would have provoked U.S. retaliation. Second, with the Arabs backing out on the PLO, it did not behoove the Kremlin to plunge itself in an Arab war. Third, its involvement in Afghanistan and Poland has greatly constrained its capacity to act on an uncertain front. Moreover, it might have calculated that the Arabs, after being defeated and disappointed with U.S. policy, would eventually turn to Moscow. These factors appear to have contained Moscow's response. As long as the Arabs remain divided and the United States continues to defend Israel, Moscow will probably refrain from involvement on the Arabs' behalf. Having no leverage with Israel and the moderate Arabs, who outnumber the radicals, the Soviet Union would hardly pursue an active policy.

According to Robert O. Freedman:

Moscow's position in the Arab world, despite the victory of its client Syria, had to be somewhat shaky as a result of the Lebanese crisis. In the first place, the USSR did not provide any meaningful political or military assistance to the PLO or Syrian fighting forces in Lebanon during the 1982 war or to Syria during its confrontations with the United States, and Moscow's utility as an ally to Arab world clients must have become somewhat suspect as a result.[26]

ISRAEL'S DEBACLE

The Israelis broke up a PLO "state within a state" in Lebanon that was the launch pad for guerrilla attacks on Israel. But on other counts "Israel made a fundamental miscalculation in its 1982 invasion of Lebanon and achieved none of its broader goals."[27]

The overall casualty figures for the decade-old civil war and Israeli invasion of Lebanon are difficult to pin down, but Bruce R. Kuniholm estimates that the war in Lebanon has "punctuated the loss of 100,000 lives."[28]

Israel's efforts to install a stable, friendly, Christian-led government in Beirut backfired. Instead, Israel's most implacable enemy, Syria, now holds more sway than ever over Lebanon and its beleaguered president, Gemayel.

The invasion has weakened Israel's international standing, and led to an ineffective, costly U.S. involvement in Lebanon. Three years of Israeli occupation turned Lebanon's Shiites into a powerful anti-Israeli fighting force.

When Israel moved into southern Lebanon in 1982, it said it was seeking to establish a 25-mile buffer zone against Palestinian border attacks. But the assault force quickly moved 60 miles to Beirut, pounded the western, PLO-held half of the city in a two-month siege, drove the Palestinians from the city, and oversaw the installation of Bashir Gemayel, Christian militia warlord, as president.

The Israelis hoped for a lasting treaty with a pacified Lebanon. But Bashir Gemayel was assassinated on September 14, 1982. Two days later, vengeful Christian militiamen massacred hundreds of Palestinians and Lebanese Muslims in refugee camps in Israeli-controlled Beirut. Peace was not at hand. Soon after the massacre, as the world condemned the slaughter, the Israelis began to pull back from central Lebanon, leaving a vacuum in which new fighting erupted among Lebanon's communal factions.

The United States—backed by France, Britain, and Italy—were drawn into the Lebanese quagmire as "peacekeepers," trying to prop up the shaky Christian-led government. But, like Israel, the United States was frustrated by the jigsaw puzzle of Lebanon's deadly politics. Despite the massive supporting firepower of the U.S. Sixth Fleet, the Reagan administration was forced to withdraw the U.S. Marines from Lebanon after suicide bombers blew up the U.S. Embassy and the Marine Corps headquarters, killing some 300 people.

On March 5, 1984, a week after the Marine pullout, the Lebanese government canceled its accord with Israel stipulating the promise of Israeli troop withdrawal. Later, under hit-and-run guerrilla attack, the Israelis had to speed their retreat. In the end, only the Syrians and their Lebanese allies have emerged from the carnage and confusion with anything approaching gain.

In the invasion's early days, the Israelis mauled the Syrian army in Lebanon. Syria's military machine was rebuilt by the Soviets and Assad is able to assert even more authority over Lebanon than before the Israeli invasion.

Supported by Syria, the downtrodden Shiites are ascending as a major new force. The Shiites, largest of Lebanon's religious communities, turned violently against the Israeli occupation force when Israel clamped down hard on Shiite-populated southern Lebanon in order to crush lingering anti-Israeli resistance in the area. By the end of 1985, leaders of the 6,000-member Amal militia were boasting that they had defeated the formidable Israeli army. New-found Shiite influence could tip the political scales in Lebanon as feuding religious communities struggle for power in the aftermath of the Israeli withdrawal.

NOTES

1. The current population figures for Lebanon are taken from various issues of *New York Times*, including June 18, 1985.

2. William R. Brown, "The Dying Arab Nation," *Foreign Policy* 54 (Spring 1984): 34.

3. See J. P. Spagnolo, *France and Ottoman Lebanon 1861–1914* (London: Ithaca Press, 1977), pp. 97–125.

4. For the text of the Lebanese Constitution, see Amos Jenkins Peaslee, *Constitution of Nations* (The Hague: Martinus Nijhoff, 1966), vol. 11, pt. 1, pp. 633–47.

5. Frank Stoakes, "The Civil War in Lebanon," *World Today* (London) 32, no. 1 (January 1976): 9.

6. Kamal S. Salibi, *Crossroads to Civil War: Lebanon 1958–1976* (Delmar, New York: Caravan, 1976), pp. 1–2.

7. See the description of some of the activities of the PFLP in Harald Vocke, *The Lebanese War: Its Origins and Political Dimensions* (London: C. Hurst, 1978), pp. 33–34.

8. *Newsweek*, July 12, 1982.

9. *Guardian Weekly* (London), September 12, 1982.

10. Ibid., October 3, 1982.

11. *Newsweek*, July 12, 1982.

12. Ibid.

13. See why the Arab leaders hate the PLO, in Brown, "Dying Arab Nation," pp. 31-37.

14. Jonathan C. Randall, *Going All the Way: Christian Warlords, Israeli Adventures, and the War in Lebanon* (New York: Viking Press, 1983), pp. 215-16.

15. Ibid., pp. 1-24.

16. Ibid., p. 136.

17. Ibid., p. 119.

18. Lewis S. Snider, "The Lebanese Forces: Their Origins and Role in Lebanon's Politics," *Middle East Journal* 38, no. 1 (Winter 1984): 1-33.

19. Ibid., p. 17.

20. Ibid., p. 30.

21. *Washington Post*, January 27, 1984.

22. *New York Times*, February 8, 1984.

23. Ibid., January 23, 1984.

24. *Newsweek*, September 6, 1982.

25. *Guardian Weekly* (London), September 12, 1982.

26. See Robert O. Freedman, "Moscow, Damascus and the Lebanese Crisis of 1982-1984" (Paper presented at the American Political Science Association conference in Washington, D.C., August 30, 1984).

27. Robert G. Neumann, "Assad and the Future of the Middle East," *Foreign Affairs* 62, no. 2 (Winter 1983-84): 237.

28. Bruce R. Kuniholm, "Lebanon: Two Analyses of How to Get Out of There," *Durham Morning Herald*, February 12, 1984. See also the companion article, under the same title, by Bruce B. Lawrence.

8

Palestinianism:
A Nationalism Denied

No other topic of discussion in recent years has generated so much polemic in political and academic circles as has the analysis of the Palestinians and their problems. Most of the discussion has been centered around the rise of Palestinian guerrillas and their acts of terrorism. The Palestinian struggle has been understood, particularly in the West, as an anti-Semitic one. Any scholarly attempt to understand the Palestinian issue will fail unless one reviews the history of the Palestinian people and the Jewish interest in Palestine.

It is possible to construct a case for Palestinians and Jews based on history. The Palestinian case rests on the premise that imperial powers, in conjunction with Zionists, imposed Jewish migration from Europe against the wishes of Palestine's population. The Jewish case is based on the aspirations of a people who have suffered exile and persecution. Both sides regard their right as self-evident and based on logic and historical background.

Over a period of many centuries, the Palestinians have developed deep historical roots in Palestine. The indigenous Arab population reaches back to the beginning of history. What a noted scholar has said with regard to ancient Palestine is no less applicable to Palestine during the many centuries since:

> For a hundred years or more, history has been explained as a constant
> process of barbarian invasions, of "fresh blood" invading and destroying

older populations who had become corrupt and degenerate. Historians have thus carried out more genocides than has all humanity. This old theory is simply no longer satisfactory, far too simple, and most seriously misleading.[1]

The Jews had their first contact with the "promised land" about 1800 B.C., when Abraham led his followers to the outskirts of Palestine, then controlled by the Canaanites. Later, the Jews migrated to Egypt and lived there as slaves for several centuries before Moses led them out again. The Jews returned to Palestine around the twelfth century B.C. They remained weak and divided until united by Saul. His successor, David, extended the country's borders. David's son Solomon built the first temple in Jerusalem. This kingdom lasted about two centuries before dissolving into the kingdoms of Judah and Israel. After the collapse of Israel around 700 B.C., the only part of Palestine which remained was the kingdom of Judah. It lasted until A.D. 70 as a Jewish possession.

After the fall of Judah, Palestine became a province within the Roman Empire. The Romans were succeeded by the Byzantine Empire until the seventh century. This was followed by a short Persian rule. The Muslims conquered Palestine in A.D. 633, and ruled it until the middle of the ninth century. It then fell under Egyptian control. The Egyptian caliph governed Palestine until the end of the eleventh century. For a brief period Palestine was also ruled by the Turks. The crusaders controlled Palestine for about 200 years until it was defeated by Saladin. Once again Palestine came under Muslim rule, only ending with the defeat of the empire in World War I. A British mandate was established by the League of Nations, which remained in effect until Israel was created by the United Nations in 1948.

We mention this chronology not to catalogue the obvious misfortunes of the Jewish people, who migrated from Palestine to other countries, but more to emphasize the fact that today's Palestinians are the descendants of the people who have lived there since time immemorial, who have continued to live there through a succession of conquests, and who have shared in and contributed to the culture and development of Palestine. With the Arab conquests more than thirteen centuries ago, the people of Palestine began to identify themselves as Arabs.

This Palestinian identity did not come about easily, nor is it complete. It has existed since the Romans coined the term "Palestine" in the second century A.D. But, it was the confrontation of two incompatible nationalisms--Palestinianism and Zionism--that produced the troublesome problem for Palestine.

"A land without a people for a people without a land," was the way Theodor Herzl, a European Jew, once described Palestine. But the territory the Jews had left earlier was not vacant when their descendants returned; they found Palestinian Arabs settled there. Now the Palestinians are "a people without a land."[2] The establishment of modern Jewish settlements in Palestine beginning in the 1870s stimulated Palestinian nationalism. The roots of the Jewish state can be traced to 1895 when Herzl wrote *The Jewish State*. In this book he advocated the creation of a Jewish state as a remedy to anti-Semitism in Europe. Referring to the location of the proposed state, Herzl wrote, "We shall take what is given us, and what is selected by Jewish public opinion."[3]

THE WORLD ZIONIST ORGANIZATION

In response to Herzl's call, the first Zionist congress met in Basle, Switzerland, in 1897. In his address, Herzl stated the objectives of the meeting in the following words: "We are here to lay the foundation stone of the house which is to shelter the Jewish nation."[4] To attain this objective, all Jews were to be organized in a worldwide Zionist movement, which came to be known in later years as the World Zionist Organization.

Since the diaspora (dispersion), religious Jews had cherished the messianic hope of one day returning to the Holy Land, but it was essentially a spiritual sentiment unconcerned with political ambition or possession. Now, a few Western Jews, some atheists or agnostics, sought to convert it into a powerful national-racial movement. They encountered opposition, especially from religious Jews, who believed that a return to Zion must be carried out through divine intervention, not through temporal means. They also feared that confusion of Judaism with nationality would undermine the hard-won position of the Jews as citizens and nationals of their adopted countries. Despite opposition, the militant Zionists, backed by powerful financial and political interests in Europe and America, soon arrogated to themselves the right to act as spokesmen for the world's Jewry.[5]

The World Zionist Organization, under the leadership of Dr. Chaim Weizmann, the Russian Jew who succeeded Herzl, contributed to the war efforts of the Allies during World War I. During the war, the British government, finding Turkish aid to Germany imminent (Palestine was under the Turkish rule) and U.S. aid to the Allies still doubtful, made certain promises to the Arabs and Jews. The Zionists aimed to convert Palestine into a national home for the Jewish people. The objective of the Palestinian Arabs was independence.

Through his friendship with leading British leaders and journalists, Weizmann was able to make the British government aware of the Zionist desire to create a homeland for the Jewish people in Palestine.[6] The British government was told of the support the Allies would receive from the Jewish people if it supported the Zionist program. The British government believed that a Jewish state in Palestine would strengthen the British position in the Middle East.

Until World War I, British policy in the Middle East had revolved around the maintenance of the Ottoman Empire. But when Turkey joined Germany in the war, British promises to the Middle East were shattered. Britain wanted a new order for the Middle East, in which Arab autonomy would supplant Ottoman rule. Britain asked Hussein, the grand sharif of Mecca, for an Anglo-Arab alliance against Turkey. Though eager for independence, the Arabs were held back by suspicion of European designs and an aversion to fight against Muslims. The sharif later was tempted to accept the British offer on the basis of complete independence of all Arab lands then under Turkish rule. The Anglo-Arab agreements to that effect, reached in the fall of 1915, led to the Arab revolt against Turkey in 1916.

In 1917, a new British cabinet was actively searching for ways and means to extricate itself from the agreements which its predecessor had reached with France for the postwar division of the Ottoman Empire. It was at this point that Zionist attempts to secure British support for a Jewish-dominated Palestine were reactivated.

The Zionist campaign in Great Britain was followed with a massive one in the United States, led by Supreme Court Justice Louis Brandeis and Rabbi Stephen Wise. With the help and support of prominent U.S. officials, the news media, and other important organizations, the World Zionist Organization in the United States succeeded in getting the full support of President Woodrow Wilson and other government leaders.

The campaign in the United States and Great Britain led to the Balfour Declaration on November 2, 1917, which promised a national home in Palestine for the Jewish people. The declaration stated that nothing must be done to prejudice the civil and religious rights of the existing non-Jewish communities in Palestine. The Balfour Declaration was not a legally binding document, for Palestine was still a part of the Ottoman Empire. Nevertheless, Great Britain took it upon itself to give what did not belong to it. Although the declaration carried no authority in international law, it marked the beginning of the Western conspiracy against the Palestinians.[7]

PALESTINE MANDATE

By the end of World War I, the British government sought to implement the Balfour Declaration. At San Remo in 1920, Palestine was awarded to Great Britain as a mandated territory. The former Turkish ruled Arab territories were split further into jurisdictions—Lebanon and Syria under France, and Iraq and Transjordan under Great Britain. Britain appointed Sir Herbert Samuel, an English Jew, as its first high commissioner in Palestine. Britain recognized the World Zionist Organization as a representative Jewish agency. It opened the gates of Palestine to massive Jewish immigration, despite Arab protests.

Meanwhile, the King-Crane Commission, appointed by Woodrow Wilson, raised serious opposition to the establishment of a Jewish state in Palestine. The commission reported that:

> The non-Jewish population of Palestine—nearly nine-tenths of the whole—are emphatically against the entire Zionist program .... There was no one thing upon which the population of Palestine was more agreed than upon this.[8]

With the establishment of the mandate on July 24, 1922, all of whose principal clauses were of Jewish origin, the Zionists began entering Palestine "of right," not on sufferance, on the claim that they were to build their Jewish national home. They made Hebrew an official language. "On the very postage-stamps of Palestine," noted a British writer, "they had the words *Eretz Israel* (the land of Israel) placed. But it would have been nearer the truth if the inscription had been *Ersatz Israel* (substitute Israel). They might talk Hebrew, but there was not a Hebrew deed done by them; they had, in a sense, to translate all their acts into it."[9]

The civilization they brought was thoroughly European. For the indigenous Jews, and the rest of the indigenous population, they had nothing but contempt. Insisting on total segregation from the indigenous population in every sphere of life, the European Jews proceeded—with the connivance of the mandatory power—ruthlessly to lay the foundations for an exclusivist Jewish state.

Under the mandate the Zionist slogans "Jewish Labor" and "Jewish Produce" were more vigorously applied. The principal victims of the Jewish boycott of indigenous labor were the *fellahin*, who were evicted from lands bought by Jews. Jewish land purchases—almost all from absentee non-Palestinian landlords—could only take place after tenants had been evicted. Thus, farmers whose families had lived and worked on the land for generations were deprived of their homes and livelihood. They received little compensation, if any. Non-Jews were forbidden to work on Jewish-owned land, since the Constitution of the Jewish Agency for Palestine required that land purchased should be the property of the Jewish National Fund. The fund, which eventually came into possession of 90 percent of the lands acquired by Jews, prohibited employment of non-Jewish labor on pain of fines and eviction.

Despite the provocation of Jewish money the indigenous population did not sell its land. By the end of the mandate, Jews held less than 6 percent of the total land area of Palestine and only 10–12 percent of cultivatable land. The indigenous Arab population still accounted for 70 percent of agricultural income, including a nearly 50 percent share in Palestine's citrus fruits.

The Arab population was also barred from employment in Jewish industry. The *Histadrut* (General Federation of Jewish Labor), itself excluding non-Jews, enforced this ban and at the same time exercised a decisive influence in determining both the number and the Zionist qualifications of Jewish immigrants. The aim of Jewish industry, moreover, was not to cater to Arab wants, three quarters of whom were poor farmers. The Arabs, commented J. M. N. Jeffries, "were thought of as mere livestock in the fields."[10]

The Jewish settlers had the immense advantage of vast infusions of outside capital. Capital transferred to Palestine by Zionist organizations for the benefit of Jewish settlers totaled for the period 1919–48 more than $1.5 billion. This means an average expenditure of $3,225 for each immigrant. Extensive credit was also made available to Jewish settlers. The income of the Arab farmer at the same time was barely 40 percent of that of the Jewish farmer.

The Jewish boycott of Arab products combined with the heavy inflow of Zionist capital and the Zionist-British control of the banking system blocked Arab industrial development and the emergence of an Arab capitalist class.[11] By the early 1930s the indigenous population accounted for at least 40 percent of the income from trade and services, but its industrial system accounted for only 10 percent of the country's industrial production. The lag in capitalist development meant that uprooted farmers could find no alternative work and the frustrated Palestinian intelligentsia could find no useful employment for its skills.

By the mid-1930s thousands of landless and jobless farmers were crowded in shacks on the edges of cities like Haifa. With the rise of Jewish immigration in 1935, Palestinian unemployment increased. In Jaffa alone 5,000 were unemployed, yet the mandate government continued to issue permits for Zionist workers and favored such workers in distributing what jobs there were. In the wholly Arab city of Jaffa, for example, the contract for building a public school went to a Jewish contractor employing only Jewish labor.

HEALTH, EDUCATION, AND WELFARE

Jewish-British policy revealed all the characteristic features of colonial oppression. Laying the groundwork for their future state, the Zionists established their own school system, hospitals, pharmacies, banks, credit cooperatives, and military forces. The indigenous population was excluded from all of these. These institutions assumed a racist character and all the more effectively embodied the embryo of the future exclusivist Jewish state.

The mandatory power cooperated fully in the Zionist attempt to put the indigenous population in a position of ever greater inferiority. It granted the Jews exclusive rights to exploit Palestine's resources. It spent 40 percent of its budget maintaining law and order, mainly to crush the increasingly active opposition of the indigenous majority. Most important, it held its public social welfare, educational, and economic development expenditures to a minimum.

In 1939, for example, government health expenditures for both Jews and Arabs totaled 243,016 Palestinian pounds, or on a per-capita basis, 88 U.S. cents. In the same year total health expenditure by Jewish institutions (for Jews only) totaled 525,150 pounds or $5.61 per Jewish inhabitant. It is no wonder that the incidence of trachoma among Arab school children (49 percent) was far greater than among Jewish school children (2.4 percent). The Jewish minority got 50 percent of hospital care. Its Jewish hospitals took in only a negligible number of non-Jewish people, although non-Jewish hospitals treated a substantial number of Jewish people.

Discrimination in education was even greater. Up to 80 percent of the large network of schools and technical and agricultural institutes established by Jewish organizations for Jews only were financed by the government. Public education provided for the Arab inhabitants of Palestine was "on an even more primitive level than its health service," according to a Jewish economic study in 1948.[12] In the 850 Arab villages and towns of Palestine, there were in 1941–42 only 404 Arab public schools. Only 20–25 percent of 5- to 15-year-old Arab children could attend school. This at a time when 90 percent of Jewish children of the same age group were in school.

The Jewish population of less than half a million had 62,000 children in Jewish community schools and 2,827 teachers in 1941. In the same year the Arab population of over a million had only 56,600 pupils in public schools and 1,456 teachers. Thus, Arab children had only one teacher for 38.8 pupils while Jewish children had one teacher for every 22.2 pupils. Arab public school buildings and equipment were of a much lesser quality than those of Jewish schools.

A similar gulf existed between the standard of public services provided by Jewish and Arab local governments. Jewish communities, enjoying the benefits of Zionist financing from abroad as well as income several times higher than those of Arab communities, spent four to eight times as much per capita on local public services as did the much poorer Arab communities.

THE INTERWAR PERIOD, 1919–39

Denied all political rights and means of self-defense, the indigenous Arab majority rebelled again and again—in 1921, 1929, 1933, and 1936–39—against the British-backed Jewish drive to take over their country and for the right to self-determination and independence. To quell the 1936–39 rebellion, the British were forced to employ 20,000 troops in addition to their regular police forces. Although the rebels were militarily defeated, they won from the British the promise of independence within ten years and a phased end to Zionist immigration within five years.

This pledge, like so many others, was not fulfilled. Deprived of social, political, and economic development by four decades of deliberate discrimination, and weakened by the loss and expulsion of their most promising leaders after the 1936–39 rising, the Arab people of Palestine fell victim to the Jewish onslaught.

Britain transferred state lands to the Jews for colonization. It protected the institutions of the fledgling "national home." Britain permitted the Jewish community to run its own school and to maintain its military establishment—the *Haganah*. It trained mobile Zionist striking forces—the *Palmach*—and condoned underground terrorist organizations. During the interwar period, the World Zionist Organization began to send money into Palestine and increase Jewish immigration.[13] The Jewish population during this period increased by approximately 30 percent of the total population. By the mid-1930s a British commission had to come to describe Palestine as a "state within a state."

TABLE 8.1. Population and Immigration (in thousands)

Year	Non-Jewish Population	Jews	Jewish Immigration
1914	604.3	84.7	N.A.
1919	642.0	58.0	1.8
1922	618.3	83.8	8.7
1931	851.7	172.0	4.1
1935	940.8	320.4	66.5
1939	1,056.3	445.5	31.2
1946	1,237.3	608.2	18.8

Source: Palestine Royal Commission Memorandum, Prepared by the Government of Palestine, A Summary of Palestine (1946), UN Special Commission on Palestine, Report of General Assembly 1947.

To head off Arab concern, the British government assured the Arabs that their rights would not be prejudiced by the rapid growth of the Jewish community. The Arabs were denied analogous facilities and deprived of the means of self-protection. After 30 years of British rule, the Jewish community had grown to 12 times its 1917 size and accounted for one-third of Palestine's population (see Table 8.1).

Arab resistance to British policy mounted. Mass demonstrations and strikes were conducted by the Arabs, causing the British government to send a number of commissions to investigate the situation. All commissions recommended restriction of Jewish immigration to Palestine, and stressed that the Jews had no rightful claim to Palestine. However, in 1939, the British government issued the White Paper, which laid down principles calling for the continuation of British rule in Palestine for a period of not less than ten years; an allowance of 75,000 new Jewish immigrants; and the lifting of restrictions on land sales to the Jews. The Jews rejected the document and called for the creation of a Jewish state in Palestine. Arab reaction to the British document varied. On the whole the Arabs also rejected the White Paper and demanded that Jewish immigration be stopped.[14]

By the end of World War II, the growing opposition of newly emerging nations, including Arab states, to Britain's role in Palestine had forced Britain to exercise some restraint in its wholehearted support for the Jewish cause. The rise of the United States as a superpower, with economic and strategic interests in the Middle East, and the growing responsiveness of U.S. politicians to the Jewish cause offered Jews the prospect of alternative support for Israel. Along with the U.S. government, Jewish Americans offered the settlers powerful and militant support to see it through the forthcoming struggle for statehood.

When the Jewish holocaust was uncovered at the end of World War II, Zionist organizations capitalized on the West's sympathy. When the Labor party won the election in Britain, the hopes of the Jewish people were raised, because the Labor party had been a champion of the Jewish cause. In the meantime, the World Zionist Organization moved its base from Europe to the United States. It held its first congress in New York in 1942. The Jews called for increased immigration of their people into Palestine and the creation of the state of Israel.[15]

In the mid-1940s, Jewish colonizers of Palestine, sheltered and nursed for 30 years by British imperialism, were ready to look for a more powerful supporter. The United States was a willing partner that admirably fit their requirements.

ISRAEL IS CREATED

The League of Nations was the instrument to bestow on the Anglo-Zionist partnership a semblance of international respectability. The United Nations was selected for a similar purpose by the U.S.-Zionist *entente*. Britain had prevailed upon the predominantly European League of Nations to endorse a program of European Zionist colonization in Palestine. The United States led the European-U.S. majority in the United Nations to overrule the opposition of the Afro-Asian minority in the General Assembly, and to endorse the establishment of a colonial Zionist state in the Afro-Asian bridge, the Arab land of Palestine. Apart from South Africa, no Asian or African nation favored the partition plan for Palestine proposed to the General Assembly by its special committee on Palestine. Although in the final vote on November 29, 1947, one Asian country and one African country other than South Africa did vote for the adoption of the recommendation, enthusiastic support for the proposal came exclusively from Europe, Australia, and the Western Hemisphere.

On May 14, 1948, the British mandate was formally terminated and the same day the state of Israel was proclaimed. Just 11 minutes after the proclamation of statehood, a *de facto* recognition was granted to Israel by the United States.[16] Soon afterward the Arab armies of neighboring countries entered Palestine in defense of the Palestinians. For a brief period at the start of the war, the Arabs gained the upper hand, but the big powers quickly intervened to impose a United Nations-sponsored truce. When the fighting resumed the Arabs found that the Israelis had used the interval to stock modern weapons from various sources, while Britain in observance of the United Nations arms embargo, effectively cut off the armies of Iraq, Egypt, and Jordan from their only source of arms.

It was at that stage that the Palestinian Arabs, debilitated by 30 years of British suppression, proved incapable of withdrawing their assault on the Jewish community, organized, armed, trained, and supported by the European, U.S., and international community.[17]

The Palestinians lost not only the battle for political control of their country—they lost the country as well. Palestinians were forcibly expelled from their homeland, thus ruthlessly removed from their rightful lands. Palestine was opened for a well-organized and liberally financed new wave of colonization, speedily executed in order to create a *fait accompli*.

As soon as it was safely entrenched, Israel embarked on its preplanned course of expansionism in line with the declaration of its leaders that the ultimate borders of Israel would stretch from the Nile to the Euphrates. Israel refused to allow the refugees to return home and not only clung to the territory occupied during the war and just after the armistice, but also occupied the demilitarized zone, driving out thousands of Arabs.

But, for all the means of survival it acquires, most recently from the United States, the Zionist state remains an alien body in the region. This is because of its vital and continuing association with Western imperialism, its introduction of colonialism to Palestine, and its pattern of racial exclusiveness and segregation. No words could better describe the essentially alien character of the Zionist state than the following passage, written by the first Israeli prime minister, David Ben-Gurion:

> The state of Israel is a part of the Middle East only in geography, which is, in the main, a static element. From the decisive aspects of dynamism, creation and growth, Israel is a part of world Jewry. From that Jewry it will draw all the strength and the development of the Land; through the might of world Jewry it will be built and built again.[18]

ZIONISM: EXCLUSIVIST NATIONALISM

Zionism, identifying religion with race and a "chosen people," is characterized by concepts of Jewish exclusiveness and uniqueness. It is, further, an ideology rooted in Europe. Zionism is consequently antithetical to the pluralist traditions and open collective mentality of many people of the Middle East as well as being without roots in the cultural environment of the area. From the beginning the Zionist aim was to transform Palestine into a state from which all non-Jews would be excluded, a new ghetto where Zionists could subject to its discipline the entire Jewish people. The assumption of the nonassimilability of the Jews led Zionists to reject emancipation (the integration of the Jews as individual citizens with equal rights and obligations in the nations in which they live) and to try to reestablish in one place the isolated exclusivist ghetto communities of the middle ages.

From the beginning Zionist settlers isolated themselves from the indigenous population whose very existence they long denied. To the Zionists the Palestinians were nonpeople. They simply considered the natives unworthy of serious consideration. So complete was the rejection of any association with the native Palestinians that—unlike other settler movements—Zionists did not want the natives even as a source of cheap labor. Through Jewish labor the Zionists asserted racial supremacy over the indigenous population. Their belief in the Jews as the "chosen people" went beyond—and was potentially more brutal than—the normal sense of European superiority over "backward natives" common at the time. Yet the Zionist aim was exactly the same as that of all settler minorities: to counteract the unfavorable population ratio by imposing and maintaining social, economic, and technological supremacy.

The Zionist division of mankind into Jews and gentiles found clear expression in the Balfour Declaration which was written with meticulous care by the Zionists themselves. The declaration refers to the indigenous population as "non-Jewish communities." These so-called non-Jewish communities were Arabs, constituting 88 percent of Palestine at that time, and were virtually unanimously opposed to the Zionist intrusion into Palestine. Sir Louis Bols, Britain's chief political officer in Palestine, stated "that approximately 90 per cent of the population of Palestine is deeply anti-Zionist. This opposition comprises all Moslems and Christians and a not inconsiderable proportion of Jews."[19]

THE JEWISH STATE

In November 1948 the Israeli census counted only 130,000 Arab Palestinians, that is, only 15 percent of the original Palestinian Arab population. Such an expulsion of almost an entire indigenous population by an alien settler movement in a period of hardly one year is unprecedented.

The new state stubbornly refused to heed the constantly repeated United Nations resolutions calling for the repatriation of the displaced Palestinians. Instead it enacted the Law of Return which allowed any Jew in the world to become a citizen upon his arrival in the country. By this law the qualifications of the would-be citizen are determined by parentage according to a definition set millennia ago. Under the Law of Return, 1,290,771 Jews immigrated to Israel between May 5, 1948 and December 31, 1948.

The Israeli nationality law gives Jews citizenship by right, but the Arab Palestinians who have not fled may become citizens only by grace: they must furnish proof and their applications may be refused without explanation. An estimated 60,000–70,000 Arabs live within Israel today but about 20 percent of them are stateless. Their number is growing because statelessness is inherited. Children born to parents without citizenship, who may be unaware of their status until they apply for passports or other documents, are also stateless though they have been born in Israel, where their families have lived for generations. A stateless Arab who wants to leave the country will be given a travel document valid for one year and a day; if he does not return within the year, he will be barred forever. As Aharon Cohen has observed:

> ... the Arab minority in Israel lives in condition of painful national discrimination. ... and the fact that any Jew arriving in Israel can receive citizenship automatically, whereas an Arab born in Israel and living there all his life is asked to prove his right to it with documents and witnesses.[20]

Nowhere in Asia and Africa—not even in South Africa—has European racism and supremacism expressed itself with such a zeal for racial exclusiveness and expulsion of native populations from the settler-state as it has in Palestine. Perhaps this divergence of Zionism from the norm of European colonization may be explained by the fact that dedication to the racist doctrines inherent in the ideology of Zionism has preceded, stimulated, inspired, and at every stage guided the process of Zionist colonization in Palestine since the inauguration of the new Zionist movement of the nineteenth century.

So long as they were powerless to dislodge the indigenous Arabs of Palestine, Zionist colonists were content with isolating themselves from the Arab community and instituting a systematic boycott of Arab produce and labor. Accordingly, from the earliest days of Zionist colonization only Jewish labor could be employed in Zionist colonies. The Jewish Agency, the Jewish National Fund, the Palestine

Foundation Fund, and the Jewish Federation of Labor vigilantly ensured the observance of that fundamental principle of Zionist colonization.

As early as 1895, Theodor Herzl was busy devising a plan to "spirit the penniless population across the frontier by denying it employment."[21] In 1919, Chaim Weizmann was forecasting the creation of a Palestine that would be "as Jewish as England is English."[22]

The Zionist concept of the final solution to the Arab problem in Palestine and the Nazi concept of the final solution to the Jewish problem in Germany consist of the same basic ingredient: the elimination of the unwanted human element in question. The creation of a Jew-free Germany was sought by the Nazis through more ruthless and more inhuman methods than was the creation of an Arab-free Palestine accomplished by the Zionists. But behind the difference in technique are identical goals.

The fate of Palestine represents an anomaly, a departure from the trend in contemporary world history. Scores of nations were enjoying their right to self-determination at the same time the Arab people of Palestine were finding themselves helpless against the systematic colonization of Palestine by foreigners. This climactic development took the combined form of forcible dispossession of the indigenous population, their expulsion from their own country, the inplantation of an alien government on their soil, and the speedy importation of hordes of aliens to occupy the land emptied of its rightful inhabitants.

The Palestinians have lost not only political control over their country, but its physical occupation as well. They have been deprived of their inalienable right to self-determination and their elemental right to exist on their own land. This dual tragedy, which befell the Palestinians in the middle of the twentieth century, symbolized the birth of Palestinianism.

INGATHERING AND LAND ALIENATION

A unique feature of Israel has been the unlimited immigration—the "ingathering"—of Jews from all over the world. David Ben-Gurion considered this ideology as the *sine qua non* of Zionist fulfillment.[23] Parallel with the ingathering of new settlers, the state of Israel pursued an unremitting effort to reduce both the numbers and the status of the indigenous Palestinians remaining in the areas under its control and to seize by any methods the land owned by the remaining Palestinians.

Forcible expulsion of the indigenous population over the frontiers and to other parts of Israel coupled with the seizure of their lands and homes have continued from the cessation of hostilities until the present.

The government has given itself extensive powers to seize land and properties owned by Palestinian Arabs within Israel. In 1948 and after, this policy was particularly embodied in seven laws created to legitimize the theft of the lands and properties of the Arab population expelled in 1948, provide a legal basis for further confiscation from the Palestinian Arabs within the state, and deprive the Palestinian Arabs of their properties and means of livelihood and compel them to leave.

Under these laws at least one million *dunums* of land (one *dunum* is a tenth of a hectare) were stolen from the Palestinians who remained in Israel. Not a single Arab village escaped. Arab agriculture suffered heavy damage. Two thirds of the land left in Palestinian hands was rocky or mountainous and difficult to cultivate. Arab agriculture was deliberately shackled by state measures to consolidate and expand Jewish agriculture at Arab expense. No opportunities to learn new skills were offered to the uprooted farmers and agricultural workers who had to leave their villages for towns where they could get only the most menial or backbreaking jobs. Most Palestinian landowners rejected the derisory compensation offered them.

Among the most important and cruel of these laws was the absentee property law, which opened the way to the seizure of hundreds of thousands of *dunums* of land and other properties from Arabs considered to be Israeli citizens but held to be absentees. An absentee is any citizen who left his usual place of residence between November 29, 1947, and the date when the state of emergency is abrogated (it never has been) for any length of time either to visit or live outside Palestine or in any place in Palestine not then under Jewish rule. The law, which does not mention the word Arab, has been applied only to Arabs. A simple written declaration by the custodian of absentee property makes a citizen an absentee. The custodian, moreover, cannot be questioned about the sources of knowledge on which his decision is based.

Under this law, lands and properties were confiscated from owners who moved for a few days from one street to another or from one village to another—as many did in search of temporary refuge during the fighting and the turmoil and confusion that followed. Under a special clause, lands were also confiscated from Arabs who visited neighboring countries or an area in Palestine not under Zionist or Israeli control before September 1, 1948. The significance of this date is that large areas of Galilee and the Little Triangle were occupied or annexed by Israel only after that date. The inhabitants of these areas often visited neighboring countries on business. Under this provision the entire population of the Little Triangle was declared absentee and the greater part of its land confiscated. Don Peretz, who estimates that 40 percent of the land owned by Arab residents was confiscated by the authorities as part of the absentee property policy, comments further on the situation:

> Every Arab in Palestine who had left his town or village after November 29, 1947, was liable to be classified as an absentee under the regulations. All Arabs who held property in the New City of Acre, regardless of the fact that they may never have traveled farther than the few meters to the Old City, were classified as absentees. The 30,000 Arabs who fled from one place to another within Israel, but who never left the country, were also liable to have their property declared absentee.[24]

This law also applied to the immensely valuable Islamic *waqf* property (considered in Islamic law to be the property of God to be used for the benefit of the Muslim community). *Waqf* property included one-sixteenth of the area of Palestine. In cities like Jaffa and Acre it included 70 percent of all shops. The law was also enforced against Arab properties in mixed towns where the majority of the population was forced to change its residence. In practice, all Arab property in these towns was regarded as absentee unless the contrary could be proved.

The Law of Abandoned Property, enacted in March 1950, enabled the authorities to expropriate Arab property, including the *waqf*, which included many Muslim lots and buildings.[25]

The new state controlled—without payment or compensation—all lands and properties of the 85–90 percent of the indigenous Palestinian people who had been driven out. On those who remained it placed the burden of proof to secure release of their land from the custodian of absentee property. Since the custodian's policy was one of deliberate and repeated delays, this procedure could consume several years of the petitioner's time and require much running about from one office to another. The outcome, more often than not, was the expropriation of the owner's land.

All land confiscated from the evicted Arab population and from absentees—along with all government land—was ceded to the Jewish National Fund (JNF). Non-Jews cannot buy, rent, or work on the 92.5 percent of Israeli land now controlled by the JNF. Moreover, Jews who rent land from the JNF are prohibited from leasing to or employing Arabs. In August 1967, to combat the growing employment of Arab agricultural workers by Jews, the agricultural settlement law tightened the racial clauses of the JNF constitution. Legal action was taken against Jews who tried to evade these restrictions.[26]

Arabs are thus robbed of their lands and forbidden to work on or rent Jewish-owned land. These rules were upheld in Israeli civil courts by the Law of Contract. Arabs are barred from living or opening businesses in many Jewish towns—especially in upper Nazareth and Carmiel, a town founded in 1965 on confiscated Arab land. On the seizure of land, we reproduce below an atypical Israeli ruling:

> As the courts have repeatedly held, Israel is not the state of its citizens, in the Western sense, but rather the "sovereign State of the Jewish people." The legal and institutional structure of the state, as well as administrative practice, reflect this fundamental commitment to discrimination—what we would call "racism" in discussing any other society. For the Jewish majority, Israel is indeed a democracy on the Western model, but Arabs are second-class citizens at best, in principle. Furthermore, apart from a few courageous individuals, there is little protest in Israel over the basic commitment to Jewish dominance, that is repression of the Arab minority.[27]

PALESTINIAN PEOPLE

Who are the Palestinian people? What has their historical experience been in the twentieth century, and why have they resorted to armed resistance recently? These are a few of the most important contemporary questions. Their importance stems from two basic facts. The first is that a just and lasting peace in the Middle East cannot be achieved without the Palestinians. The second and more important is that the Western world has not been aware of the answers to these questions and of the fact that the Palestinians are the cornerstone to the prolonged Arab-Israeli conflict.

At the international level, the Palestinians have gained full recognition. At a number of important conferences such as the nonaligned conference in 1973 and the Islamic summit conference in 1974, resolutions were passed reaffirming commitment to the struggle of the Palestinians for self-determination. The majority of the Third World nations also expressed their support for the PLO at the nonaligned conference in 1975.

At the United Nations General Assembly, a number of resolutions were passed calling for the right of self-determination for the Palestinians. The General Assembly recognized the PLO as the legitimate representative of the Palestinians and invited it to participate in the debates of the General Assembly and the Security Council.

BACKGROUND

About 80 percent of the original population driven out of Palestine were, according to United Nations figures, farmers and unskilled workers and their dependents, the rest being professionals, businessmen, property owners, and skilled workers. Displaced farmers and unskilled workers endured great hardships in host countries, which were poor themselves and at a stage of development where they were already overburdened with an excess of agricultural and unskilled workers. Many displaced Palestinians therefore could not find employment. Conditions in refugee camps were and are harsh, humiliating, and in some host countries restrictive.

Yet even at this level the displaced Palestinians made some contributions to the countries where they found refuge. Lebanon, for example, owes its development of citrus fruit production to the displaced Palestinians. Exile brought the largely rural Palestinians into a more urban environment. The proportion engaged in agriculture and fishing declined while those employed in the services and building industries increased. Yet skilled and specialized workers remain very few.

In 1974 about 18 percent of the Palestinian people lived in 63 refugee camps in the West Bank, Gaza, Jordan, Lebanon, and Syria. More than half of the 628,537 people living in camps are concentrated in Gaza and Jordan. These people represented roughly 40 percent of the 1.5 million displaced Palestinians registered with the United Nations Relief and Works Agency (UNRWA). The actual number in the camps is surely higher than the official figure. For example, the camp population in Lebanon in 1971 numbered 96,000 according to UNRWA statistics, and 106,440 according to a survey made by the Lebanese Department of Planning.

People in the camps suffer from severe overcrowding. The camps have expanded little in area but their population has more than doubled over the past quarter century. The Lebanese survey showed six or seven people inhabiting houses with two rooms or less and 88.5 percent of the camp population living in houses with a total living space of less than 80 square meters.

A large number of those living in refugee camps have no permanent employment. In the Lebanese camps almost 20 percent of the people have no gainful employment at all. Unemployment among women and girls is strikingly high: 96.8 percent. Owing to the low level of education and the lack of vocational and technical training, workers from the camps usually get only the lowest paying jobs. The wage earner must often support a family of eight or more. Many in the camps barely achieve a subsistence level.

In these circumstances health standards are low. Overcrowding, a diet insufficient in quantity and nutrition, and wretched living conditions—sometimes including open sewage canals and dumps inside the camps—contribute to the prevalence of dysentery, intestinal diseases, influenza, chicken pox, and malnutrition. Stomach and intestinal parasites afflict 20 percent of the children in the camps.

In 1974, UNRWA provided aid to 830,000 of the 1,583,646 refugees on its register. This aid is equivalent to ten cents per person per day. Of this, five cents provides dry rations (mainly flour) amounting to 1,500–1,600 calories per day; four cents goes toward education; and one cent goes toward health services. UNRWA medical care has helped reduce infant mortality and has brought some health improvements to the camps, but continuing poverty and deprivation mean that progress is insignificant. The inhabitants' response to UNRWA's attempts to provide preventive medical examinations and treatment has been largely negative owing to suspicion of the agency's aims. The Palestine Red Crescent Society and other Palestinian health organizations are more successful since they win the full participation of the people.

The number of Palestinians living in camps has grown over the years. Natural increase accounts for only a part of the rising camp population. The free education and health facilities provided in the camps are also a factor. Even more so is family and village solidarity. Camp inhabitants, still firmly attached to their villages in Palestine, divide the camps into sections, each section bearing the name of the Palestinian village from which they originated. Village ties draw others from the outside to join friends and relatives within the camps.

Many Palestinian villages that the Israelis razed still exist as coherent social units in the refugee camps. Family, village, and regional ties have not weakened. Social consciousness is perhaps greater. The camps developed not in accordance with some United Nations plan, but as spontaneous social formations in which the sense of being Palestinian burns as fiercely as ever. The sense of solidarity, a shared tradition, and a common destiny has permitted the dispossessed to stand up to the long years of hardship and humiliation and forge the bonds of a tightly knit and highly conscious community. To this extraordinarily tenacious social factor, the arrival of the armed *fedayeen* added a growing national consciousness. For the first time the people in the camps accepted as their leaders persons not from their own villages or regions.

PALESTINIAN CONSCIOUSNESS

One of the most significant developments so far in the Arab-Israeli conflict has been the emergence of Palestinian guerrillas. They have begun to draw world attention to their long-standing grievances, not as guerrillas and refugees, but as a political community with hopes and aspirations. Slowly and gradually a new concept of Palestinian nationhood has developed. The movement is taking firm root throughout the world. In the political arena, a sense of Palestinianism has reemerged after more than three decades. It draws its cohesion from the loss of the homeland. This attachment is deep among the peasantry, the largest component of Palestinian refugees. Life as landless farmers has sharpened their attachment.

The factor that has crystallized the Palestinian consciousness is the repeated setbacks the Arab governments suffered at the hands of Israel. Another factor that contributed to Palestinian awakening is that the bulk of the refugee peasantry, deprived of livelihood, exists on international dole, mostly in refugee camps. These refugee camps have turned out to be "political schools" and function as hothouses of politics.

The right to national liberation is an extension of the right to national self-defense, which the charter of the United Nations not only upholds but also declares to be inherent.[28] Exercise of the right of national liberation is not confined to situations in which alien domination subjects a people to its control, or in which the resources of one people are selfishly exploited by another. Exercise of the right to national liberation extends also to those situations in which the land of one people is subjected to the control of another while it is forcibly emptied of its rightful inhabitants.

Palestine subsumes all of these elements. Palestine is under foreign rule. Its resources are exploited by others. Its people are in exile. The remnants of its Arab inhabitants languish under a regime of discrimination and oppression as harsh as any supremacist regime in the world. All of this has been accomplished by violence.

The response of the Palestinians to the menace of occupation has passed through many stages. First, when the Zionists were coming in relatively small numbers and emphasizing the religious and humanitarian motives of their enterprise, while concealing the political and ideological character of their movement, the Palestinians believed the immigrants to be pilgrims animated by religious longing for the Holy Land, or else refugees fleeing persecution in Eastern Europe and seeking safety in Palestine. Second, when the second wave of Zionist colonization began to roll onto the shores of Palestine in 1877, Arab friendliness began to give way to suspicion and resentment. The removal of Arab farmers and laborers from the new Zionist colonies, and the boycott of Arab produce, aroused Arab resentment. Since Zionist colonization was still modest, the hostility it provoked remained more or less local. Third, the alliance of British imperialism and Zionist colonialism, expressed in the Balfour Declaration, opened Arab eyes to what was happening and brought home the realization that Zionists wanted nothing less than the removal of the Arabs.

The disquiet following the publication of the Balfour Declaration was momentarily calmed by a 1918 British declaration assuring the Arabs that, as far as the territories occupied by the Allied armies were concerned, "the future government of those territories should be based on the principle of the consent of the governed."[29] Only four days before Armistice Day, a widely publicized Anglo-French declaration notified the Arabs of Palestine, Syria, and Iraq that the two Allies intended "to further and assist in the setting up of indigenous governments" and "to recognize them as soon as they are actually set up."[30] These declarations soon proved to be insincere. In the meantime they served to allay Palestinian fears.

The 1919 peace conference was expected to resolve the contradictions of Allied wartime promises and inaugurate the long-awaited era of world history founded on the principle of national self-determination, on which President Wilson had spoken emphatically. These hopes dwindled and the Zionist colonists resumed after the wartime interruption. Palestinian-Arab resistance to British occupation and Zionist colonization began to grow.

The declarations of opposition were not the only Palestinian means of resistance. In March 1920, armed hostilities broke out between Arabs and Zionists in northern Palestine, and in April, Arab-Zionist fighting took place in Jerusalem. These were followed by uprisings in 1921, 1929, 1933, 1936, and 1937. These rebellions lasted until the outbreak of World War II. During the war, the Arabs cooperated with British authorities. Following the end of the war until the withdrawal of the British and the simultaneous proclamation of Israel in May 1948, Palestinian Arabs were engaged in a life-and-death battle with the British authorities as well as with the Zionist settlers.

The means by which Palestinians chose to express their opposition to the partnership of British imperialism and Zionist colonialism were not confined to declaration and rebellion. The Palestinians brought into their struggle against the Zionization of Palestine the only remaining weapon at their command: if they had no control over the immigration of the Jewish people, they could exercise some control over the sale of land to those immigrants. During the 30 years of the British mandatory administration, the Jewish population multiplied 12 times in 1917, raising the ratio of the Jewish population to one-third of the total population. Jewish land acquisition grew only about 4 percent of the total area.

When Israel was created in 1948, Palestine's Arab population numbered 1,398,000. They were in the unusual situation of possessing a high level of national consciousness without the national and political institutions to embody it. The Palestinians found themselves uprooted from their normal way of life and turned into refugees in the surrounding Arab countries. They live in exile, remembering

Palestine through personal experience and through the accounts of families, friends, and relatives.

In their determination to pursue the difficult path of national liberation, the Palestinians are encouraged by the faith in the justice of their cause repeatedly expressed by newly-liberated peoples. From Bandung to Accra, from Casablanca to Belgrade, that faith has been clearly expressed. Palestinian aspirations for the future on both political and personal levels have crystallized into a single goal: return to the homeland.

Palestinian consciousness might be high, but political organization is limited and weak. Their traditional leadership, comprising the Arab Higher Committee, led by the Mufti of Jerusalem, has been discredited. The Palestinians were not able to fill the vacuum, due to the organizational difficulties arising from their being dispersed and to their being subject to the conditions and regulations of the countries in which they had resettled.

The areas in which the Palestinians were located had a different effect on their freedom to conduct nationalist activities. A change began after the Israeli occupation of Sinai and the Gaza Strip in 1956. Palestinians were beginning to commit themselves to the liberation of Palestine. The upsurge of Palestinian feeling occurred when Israel planned to divert part of the Jordan River. President Gamal Abdel Nasser of Egypt called upon the Arab heads of state to -discuss Israeli plans for the diversion of the Jordan waters. He said "the battle of the Jordan River is part of the battle of Palestine."[31]

THE PALESTINE LIBERATION ORGANIZATION

The 1964 Arab summit conference established the Palestine Liberation Organization (PLO) as the official voice of the Palestinian community and gave it financial support. The PLO attempted to build a Palestine Liberation Army in the Gaza Strip. At the same time, the clandestine Fatah organization was emerging. Fatah emerged because its members felt that the PLO was an establishment of Egypt and other Arab governments to contain rising anti-Israeli feeling. While Fatah was expanding other groups were emerging. Now there are several dozen commando movements. Despite these fissiparous tendencies, the trend has been toward consolidation under the aegis of Fatah. Meanwhile, the PLO stands discredited because of its defeat in wars with Israel and its withdrawal from Lebanon.

Fatah, although officially founded in 1956, did not begin active operations against Israel until 1965. From then to now it has carried out provocative acts of sabotage, usually across the Syrian and Lebanese borders.

Syria, Jordan, and Lebanon are confronted with a similar problem. Tolerance of commando operations invites Israeli retaliation. But suppression of the guerrillas produces crises and civil turmoil. Egypt is the most important country to the guerrillas. They know that under present circumstances Egypt's interest lies in a diplomatic settlement with Israel. This worries the commandos because Egypt alone can determine the diplomatic terms.

In accordance with their vague theoretical stage of revolutionary growth, the commandos have succeeded in mobilizing the Palestinian community and establishing a modest infrastructure. They have also been successful in winning support from the Arab governments and promoting a more active role for the Arab armies against Israel.

The commandos are growing in number and are becoming sufficiently entrenched in the Arab states to make their eradication quite difficult. Yet the commandos are a long way from posing a threat to Israeli security.

The commandos know that the United States is committed to the sovereign existence of Israel and its territorial integrity. The commandos doubt that any U.S. government will ever recognize them. The position of the Soviets is more dangerous. It is supplying arms to some Arab governments such as Iraq, Syria, and Libya. In case of internal conflict between the Arab governments and the guerrillas, it is likely that the Soviets will stay with the status quo. In this kind of situation the commandos might get arms and training from the Chinese, who in all probability will offer arms. There are indications that the commandos are already receiving arms and training from the Chinese.

The Palestinians are developing a political community and a national consciousness. Few would contest their claims to self-determination and no one would deny that they were promised a state in Palestine. If there is to be any solution in the Middle East, it would seem that the Palestinians must be party to it. But the Palestinians, like most of the Arab states, are not willing to deal with Israel as a state. They want settlement with the Jews. The Israelis, on the other hand, have no intention of discussing issues with the Palestinians. They want settlement with the Arab states. What an unending tragedy!

NOTES

1. George Mendenhall, *The Tenth Generation* (Baltimore: Johns Hopkins University Press, 1973), p. 216.

2. The history of the Palestinians is well presented in the following two books: J. C. Hurewitz, *The Struggle for Palestine* (New York: W. W. Norton, 1950), and William B. Quandt et al., *The Politics of Palestinian Nationalism* (Berkeley: University of California Press, 1973).

3. Theodor Herzl, *The Jewish State* (New York: Scopus, 1943), p. 42.

4. Quoted in Arthur Hertzberg, ed., *The Zionist Idea: A Historical Analysis and Reader* (New York: n.p., 1959), p. 50.

5. See George Lenczowski, *The Middle East in World Affairs* (Ithaca, N.Y.: Cornell University Press, 1962), pp. 374–75.

6. Ann Williams, *Britain and France in the Middle East and North Africa* (New York: Macmillan, 1968), p. 16.

7. For an excellent discussion on the Balfour Declaration and other aspects of Palestinian nationalism, see Alan R. Taylor, *Prelude to Israel* (Beirut, Lebanon: Institute of Palestine Studies, 1970).

8. See George Antonius, *The Arab Awakening* (Beirut, Lebanon: Khayats, 1955), p. 449.

9. J. M. N. Jeffries, *Palestine: The Reality* (London: Longmans, Green, 1939), p. 413.

10. Ibid., p. 623.

11. Two of the 27 banks in Palestine were Arab.

12. Robert R. Nathan et al., *Palestine: Problem and Promise* (Washington, D.C.: Public Affairs Press, 1946), p. 350.

13. Taylor, *Prelude to Israel*, pp. 89–107.

14. Fred J. Khouri, *The Arab-Israeli Dilemma* (Syracuse, N.Y.: Syracuse University Press, 1968), p. 27.

15. David Ben-Gurion, *Israel: A Personal History* (New York: Funk and Wagnalls, 1971), p. 54.

16. Taylor, *Prelude to Israel*, p. 106.

17. For an account of the Palestine Partition Resolution and United Nations plot, see Hurewitz, *The Struggle for Palestine*, pp. 299–331.

18. David Ben-Gurion, *Rebirth and Destiny of Israel* (New York: Philosophical Library, 1954), p. 489.

19. Quoted in Jeffries, *Palestine*, p. 333.

20. Aharon Cohen, *Israel and the Arab World* (Boston: Beacon Press, 1970), p. 332.

21. Theodor Herzl, *Complete Diaries*, vol. 1, 1960, p. 88 (entry of June 12, 1895), quoted in Erskine B. Childers, "Palestine: The Broken Triangle," *Journal of International Affairs* 19, no. 1 (1965): 93.

22. Chaim Weizmann, *Trial and Error* (New York: Harper and Row, 1949), p. 244.

23. See Taylor, *Prelude to Israel*, p. 110.

24. Don Peretz, *Israel and the Palestine Arabs* (Washington, D.C.: Middle East Institute, 1958), p. 152.

25. Cohen, *Israel*, p. 332.

26. Sabri Jiryis, "Recent Knesset Legislation and the Arabs in Israel," *Journal of Palestine Studies* 1, no. 1 (1971): 53–67.

27. Cited in Elia T. Zureik, *The Palestinians in Israel: A Study in Internal Colonialism* (London: Routledge and Kegan Paul, 1979), p. 118.

28. United Nations, *Charter of the United Nations and Statute of the International Court of Justice* (New York: n.d.).

29. Antonius, *The Arab Awakening*, p. 449.

30. Ibid., pp. 435–36.

31. Gamal Abdel Nasser, *Speeches and Press Interviews, January–December 1963* (Cairo, Egypt: UAR Information Department, 1964).

9

Middle East in Superpower Strategy

The Middle East has always been coveted by outside powers. For the United States the region is important because of economic, strategic, and political factors. For the Soviet Union geographic proximity, economic interests, and ideological considerations have made the Middle East an area of great concern. The region also presents the superpowers with a large set of conflicts and problems. They are: the Arab-Israeli conflict, the Palestinian question, Egyptian-Israeli relations, the Lebanese civil war, the Iran-Iraq war, United States-Iranian nonrelations, the Soviet invasion of Afghanistan, arms supplies, access to oil, and a host of other issues.

The Middle East is not only important to the outside world because of its strategic location, but because of its oil. The Middle East possesses about 60 percent of the world's oil reserves. At different times, 20 percent of U.S. oil imports, 56 percent of Western Europe's, and 65 percent of Japan's have come from the Middle East.[1] Knowing this, both superpowers are concerned over developments in the Middle East.

Today the Middle East seems to be the center attraction, whether we talk of superpowers, former colonial powers, or other nations. The main contenders on the scene are the United States and the Soviet Union, both using every conceivable strategy to outdo the other. Although the two superpowers' involvement in the Middle East has fluctuated since the mid-1970s, the two countries have expanded significantly their respective roles in the region.

U.S. OBJECTIVES

The United States views the Middle East in terms of how the region fits into the framework of strategic problems: (1) the increasingly vigorous and strident efforts by Arab nations to gain more political and economic power from foreign control and from foreign-controlled indigenous elites; and (2) the conflict between the Arab states and Israel.

The common thread running through U.S. responses to Arab demands is the suppression of revolution and the channelling of energies of reform-oriented governments along paths acceptable to Washington, or to overthrow them if they appear to be susceptible to communist influence.

The Arab-Israeli conflict poses at least two serious problems for the United States in the context of its global strategy of containing Soviet expansion. First, as long as the conflict remains unsettled, it is difficult to organize the Arab states into any collective pact and integrate their resources and installations into the U.S. defense position. Second, continued conflict increases the risk of direct confrontation between Washington and Moscow in the Middle East as long as the United States is totally committed to guaranteeing Israel's security. As long as no settlement is close to realization, it is possible that the Arabs will seek support from other outside powers for their cause in order to reverse the effects of past defeats.

The United States has responded to the Arab-Israeli conflict as a problem relating to the cold war and United States-Soviet détente. The attention paid to it and the urgency with which the United States addresses itself to this conflict are directly related to how these flareups impinge upon East-West relations.

The U.S. attempts to keep Arab-Israeli conflict from threatening its strategic interests in the region have alternated between trying to contain or suppress the Arab-Israeli conflict, while seeking to isolate from it other issues like access to oil and security arrangements; and attempting to settle the conflict, thereby removing a major obstacle to Western defense of the Middle East against Soviet encroachment.

The selective manipulation of arms transfers to Middle Eastern recipients reflects both U.S. concern for defense against Soviet encroachment and the desire for a settlement of the Arab-Israeli conflict with terms favorable to Israel. The use of this policy instrument in both conflict arenas points out the contradictory nature of U.S. objectives in the area. United States attempts to contain, suppress, or manipulate the Arab-Israeli conflict have at times produced the very consequences that U.S. Middle East policies are supposed to check.

Under the Reagan administration, U.S. policy has shifted somewhat from the policy of the Carter years. The Reagan administration has been more categorical than its predecessor. Most important has been the revival—at least in official rhetoric—of the primacy of Soviet-U.S. confrontation. United States Middle Eastern policy once again appears to be dominated by concern over Soviet designs in the Middle East.

To a great extent, U.S. policy under Reagan has returned, after a brief respite under Carter, to an emphasis on superpower confrontation as the major motivating factor in the U.S. decision-making process.

As will be discussed in the following section, the U.S. approach in the Middle East is incoherent, vacillating, and halfhearted, while the Soviets are taking full advantage. They are consolidating their position in Afghanistan. The Soviets have a firm grip on Syria and South Yemen and are gradually establishing control over the vital maritime routes of the Red Sea and the Indian Ocean.

THE U.S. PEACE INITIATIVE

The importance of achieving a comprehensive peace settlement in the Middle East cannot be exaggerated. The Arab-Israeli War and the Iran-Iraq War are reminders of the region's volatility. Escalation of these wars could easily jeopardize the flow of oil from the Persian Gulf and draw the United States into a military confrontation with the Soviet Union. The consequence of such a confrontation would be devastating for the entire Middle East, if not for the whole world.

In this section we will examine the U.S. approach to a peace settlement in the Middle East, leading to the Camp David summit and the subsequent Egyptian-Israeli peace treaty. United States attempts to achieve a peace settlement in the Middle East have historically fallen short of their goal. The Egyptian-Israeli peace treaty was considered a major breakthrough toward reaching this goal.

Initially, U.S. policy in the Middle East centered on containment of the Soviet Union. Directly related to this policy was the maintenance of stability and peace in the Middle East as the lack of these was considered conducive to the growth of Soviet influence. Another major U.S. policy interest was the support of Israel. As détente was reached for a time between the two superpowers, U.S. policy in the Middle East began to change. "Containment and the fear of a monolithic communism have outlived their utility. The Dulles-devised regional security pacts of the fifties are remnants of a polarization that is giving way to pluralism."[2] Politically speaking, current U.S. policy in the Mideast is based on gaining the friendship of as many nations as possible and ensuring Israel's security. Of the two major principles of U.S. policy, the second interest has become more important. This is because people all over the world have come to identify the United States with Israel.

At the beginning of this century, U.S. interest in the Middle East was limited. With the British attempt to monopolize the vast oil reserves of the Middle East through the United Nations mandates, U.S. involvement in Middle Eastern affairs heightened. It claimed a share of the region's oil as compensation for having supplied the Allied powers with vast amounts of U.S. oil. The establishment of the Turkish Petroleum Company in 1928, which included several U.S. companies, was the origin of U.S. involvement in the region. Although limited at that time, the importance of this involvement grew with the enormous increase in the use of oil in the United States.

Following the end of World War I, Soviet pressure on Iran, Turkey, and Greece increased. The United States assumed the role of protector of these and other weaker states threatened by communism. The Truman Doctrine set the tone of U.S. policy in the Middle East. President Truman said: "Totalitarian regimes imposed on free people, by indirect or direct aggression, undermine the foundations of international peace and hence the security of the U.S."[3]

The most important development in U.S. policy on Israel took place in 1947, when the United States voted in the United Nations General Assembly in favor of the creation of Israel. As a result of the United Nations vote and subsequent recognition of Israel in 1948, the United States became directly involved in the Arab-Israeli conflict. Support for Israel grew as the horrors of the holocaust became increasingly known to the world and as a result of well-organized Jewish lobbying activities in the United States.

SHUTTLE DIPLOMACY

While Arab-Israeli conflicts have their own causes and characters, they are interrelated when they involve the interests of the superpowers. Apparently the Soviets do not want to destroy Israel, but force it to give up territorial ambition. This was the philosophy of the 1967 United Nations Security Council Resolution 242 that Washington and Moscow helped to draft. At one point during the fourth Arab-Israeli war, the two superpowers were about to engage in armed conflict but, thanks to U.S. strategy and the famous shuttle diplomacy of Secretary of State Kissinger, a dangerous confrontation was avoided in the spirit of détente. In the words of Nadav Safran:

> Besides defusing an explosive situation, starting off a process of negotiation for the first time in 25 years, setting up hopeful precedents for compromise, and beginning to generate mutual trust between the antagonists, the disengagement agreements involved a substantial modification of past patterns of inter-Arab politics.[4]

United Nations Security Council Resolution 338 was passed on October 22, 1973, in which Kissinger diligently incorporated provisions making Resolution 242 mandatory. Taken together, these resolutions require the parties to negotiate peace agreements "concurrently" with security arrangements. According to an interpretation by Eugene V. Rostow, a former U.S. undersecretary of state, "Under the Security Council Resolutions, Israel is not required to withdraw one inch from the territories it holds as the occupying power until its Arab neighbors have made peace."[5]

Following the end of the 1973 war, both the radical and the moderate leaders of the Arab world began to cooperate with the United States in reaching the disengagement agreements. This rapprochement placed the United States in a position of trust and influence in much of the Arab world.[6] The elements that made this change in Arab attitude possible go further than the disengagement agreements. In the words of Safran:

They include strong shared interests with the United States on the part of conservative Arab countries, beneath their expressed resentment of American policy toward Israel; strong resentment on the part of radical Arab countries toward the Soviet Union, beneath their cooperation with it; fears on the part of both radical and conservative Arab countries about the costs and outcome of continuing confrontation and war; and hopes on the part of the Arabs that American behavior in the last stage of the October War, which saved them from the humiliation of another defeat, portended a favorable change in American policy.[7]

Kissinger's shuttle diplomacy pushing the Mideast disputants toward a step-by-step settlement received a great setback following the Arab summit meeting at Rabat, which strengthened the role of the PLO in the Arab-Israeli conflict. The situation was further complicated because the region's oil-rich nations were engaged in huge arms purchasing deals and in supporting the PLO with necessary funds. According to George W. Ball, a former U.S. undersecretary of state:

> One lesson we should have learned from the experience of past months is that highly personalized diplomacy is effective only in a bilateral setting; it has limited value in a complex situation involving many countries. Thus the attempt to settle the Arab-Israeli issue by shutting out both the more activist Arab states and the Soviet Union was predestined to failure.[8]

From the beginning of his shuttle diplomacy it was clear that the Kissinger approach had been pragmatic and not legalistic. It had not been clear as to how he could secure an Israeli withdrawal within pre-1967 boundaries. George Ball, among others, believes that this could be accomplished "only in the multilateral setting of the Geneva Conference, with participation of all the principal Arab states, including the most radical, and with the Russians acting as cochairman."[9]

The implicit assumption is that Israel would be required to return to its 1967 frontiers, with only slight modifications. While this seems unacceptable to Israel, "Kissinger is understood to feel that if the United States were ready to press for a full Israeli pullback there would be no need for Soviet cooperation in the process."[10] Kissinger's rush in his diplomacy was intended to keep the momentum of negotiation going. However, the momentum stopped because of intransigence on both sides, and because of the changing role of the PLO in the settlement of the Arab-Israeli conflict. The legitimacy of the PLO cannot be denied now. Kissinger's failure to promote talks between Jordan and Israel in advance of the Rabat summit was a great opportunity missed. Previously the Arab nations demanded the destruction of Israel. While this stand was modified, they still wanted Israel to withdraw from the territories it occupied in 1967 and 1973. PLO leader Arafat reaffirmed before the United Nations General Assembly in November 1974 that the creation of a secular Palestinian state in Palestine, which includes Israel proper, was the goal for which the Palestinians were striving. So far the Arabs have fought the Palestinians' wars. Now the Palestinians themselves are emerging in a leading role in an attempt to liberate their homeland from Jewish occupation. It was precisely for this reason that the PLO was given sole responsibility at Rabat to speak and act for the Palestinians.

From the beginning of his personal step-by-step diplomatic approach, Kissinger had been warned by many analysts of the crippling contradiction inherent in his policy. Whether right or wrong, Kissinger's follow up moves and the success or failure of his diplomacy would be seen in time. He had tried repeatedly to convince Sadat of the merits of his approach to peacemaking in the Middle East, while the Syrians and the Palestinians denounced the move. The Sadat-Kissinger understanding was based on their conviction that once Israeli forces pulled back from the strategic Milta and Giddi mountain passes as well as the Abu Rudeis oil fields, peace mediation efforts in other fronts could be undertaken at the Geneva Conference. Israel agreed to the return of these territories and the oil fields in exchange for a declaration of nonbelligerency from Egypt. Sadat was not willing to make such a declaration directly, but would give explicit assurance if oil were supplied and if American civilian technicians were stationed in the Sinai Peninsula. Sadat agreed to make an implicit pledge not to use force. It was reported that the Egyptians were ready to bear the military and economic costs of nonbelligerency. But Israel wanted a categorical declaration of political nonbelligerency. Israeli reluctance is based on the ancient and still practiced concept that war is a legal means of acquiring territory, and that territory won in wartime should not be returned to its former owner without a declaration of political nonbelligerency. Egypt could probably agree to such a formal declaration if all the Arab territories occupied by Israel were returned at the same time, in other words, if the end result preceded the beginning efforts. The Egyptian attitude was that a declaration of nonbelligerency would almost be tantamount to a declaration of illegality of war by Egypt, while Israel would keep most of the occupied land in its possession. By agreeing to a declaration of nonbelligerency, Egypt would consent to the legitimization of Israeli occupation of the rest of the Arab territories.

Egypt knew that an interim settlement with Israel would arouse deep division in Arab ranks. The centrifugal forces would begin to reassert themselves in Arab politics, and Palestinian activists might resort to violence on a larger scale. As soon as the news of the last round of Kissinger's step-by-step approach became public, the Palestinians moved into open conflict with Sadat. President Assad of Syria, being fearful that a separate Egyptian settlement with Israel might diminish the prospect of getting back the Golan Heights through peaceful means, created a united political and military command with the Palestinians.

Needless to say, a peaceful settlement in the Mideast depends on complete Israeli withdrawal from the Sinai Peninsula and the Gaza Strip, including the former Egyptian outpost of Sharm el Sheikh, which overlooks the Israeli entrance to the Gulf of Aqaba and the Gulf of Suez. Sharm el Sheikh is important for Israel's entry into its only Red Sea port, Elat. Israel also insists that it must retain the Golan Heights and the Nahal outposts along the Jordan River for security reasons. Above all, Israeli annexation of Jerusalem, where the former has established many permanent settlements, poses another seemingly insoluble problem. That which is a security hazard for Israel is also one for the Arabs. They are, after all, Arab and Palestinian territories occupied by Israel.

EVENHANDEDNESS

Under these conditions, what are the prospects for a peaceful settlement? The hope is to reconvene the Geneva Conference on the basis of Resolution 242 under the auspices of the United Nations, the United States, and the Soviet Union. Arab nations, Israel, and the Palestinians are supposed to participate. The planning and organization of a conference of so many diverse elements, involving issues as complex as Israel's borders with Syria, Jordan, and Egypt, and the Palestinian problem are such that no quick decision can be expected.

Even assuming that the conference is resumed in Geneva, endless debate would build pressure on Israel to withdraw its forces from all occupied Arab territories. The U.S. policy has been one of evenhandedness. Not only are some U.S. senators and congressmen becoming more discriminating and reserved in their attitude toward Israel, even Ford and Kissinger have complained of Israeli intransigence and inflexibility.[11] On top of this Ford ordered a formal review of U.S. policy in the Mideast. The White House announcement came at the collapse of the tenth round of Kissinger's shuttle diplomacy, a time when the world was blaming Israel for its uncompromising attitude. Probably this was done to prevent the cooling of otherwise warm U.S. relations with such moderate Arab states as Egypt and Saudi Arabia. But nothing spectacular happened as a result of the reassessment of U.S. policy in the Middle East. Jewish influence on U.S. policy is meeting some opposition. At the same time an opinion favoring contact with Arabs to appreciate their point of view is appearing. Israel's problem is with the White House, not with Congress. Israeli strategy has been to stand firm, to keep up lobbying activities, and wait out U.S. reassessment of its policy toward the Mideast on the assumption that the White House cannot ignore the Israeli request for military and economic assistance unless Congress is prepared to back the president.

SPECIAL RELATIONSHIP WITH ISRAEL

President Carter reaffirmed the special relationship between the United States and Israel at a press conference on May 13, 1977, when he said:

> We have a special relationship with Israel. It's absolutely crucial that no one in our country or around the world ever doubt that our No. 1 commitment in the Middle East is to protect the right of Israel to exist, to exist permanently, and to exist in peace. It's a special relationship.[12]

As a result of this relationship the United States has been walking a diplomatic tightrope in the Mideast for the last 38 years. A tenuous relationship with Arab states, a real need for Arab oil, Third World political support, and the constant threat of Soviet penetration into the area have contributed to an urgent American desire for peace in the Middle East.

The strong preelection commitment by Carter to Israel gave way to a more moderate stance at the urging of Departments of State, Energy, and Defense. The Arab oil weapon was probably uppermost in Carter's mind.

Carter became the first U.S. head of state to discuss the concept of a Palestinian homeland.[13] He recognized the Palestinian problem as the core issue in the Arab-Israeli dispute.

The ambiguity of Carter's position on the Palestinian question added to the confusion and uncertainty of peace attempts. The imposition of a peace settlement on the Arabs and Israelis was considered impossible. Instead a joint United States-Soviet statement called for a conference in Geneva no later than December 1977 to work out a comprehensive and lasting solution to the conflict.[14] The response from the parties concerned was negative. Egypt and Israel objected to Soviet participation in the peace negotiations. Most other Arab nations objected to direct negotiations with Israel. The inclusion of a delegation of the PLO was viewed with concern by Israel and its U.S. supporters.

While the impasse continued, a dramatic development in Arab-Israeli politics took place. Anwar Sadat stunned the world by visiting Israel. This was ostensibly for the purpose of removing the psychological barriers separating the Arabs and the Israelis. The Sadat visit shifted the focus of the peace initiative from a multilateral to a bilateral one. Carter wanted to capitalize on this new development in the peace initiative and on Sadat's attitude. The latter addressed the Israeli Knesset and outlined five principles of a peace settlement. The hard-line Arab states condemned his actions while moderate Arab nations maintained a low profile.

Subsequently, Egyptian and Israeli officials met in Cairo to discuss further details of the peace proposals. Prime Minister Begin visited Washington and presented his peace proposal to Carter. His plan was designed to relinquish control of Sinai and limited self-rule for the West Bank and Gaza in exchange for normalization of relations. The Carter administration, although optimistic, showed a guarded response to the peace proposals.

In late December 1977, a meeting between Sadat and Begin was held in Ismailia, Egypt. The talks failed due to the question of autonomy of the West Bank. Further talks also resulted in a stalemate and Sadat stated that there was no hope of reaching a settlement.[15] Both sides were standing firm on their proposals. They began to criticize each other. The attempt to rescue the mood of negotiation failed and relations between the two parties deteriorated.

THE CAMP DAVID ACCORD

Carter became concerned that the impasse could jeopardize the fragile relations between Egypt and Israel and wreck any chance for peace in the Middle East. He invited the two leaders to Camp David for face-to-face talks aimed at breaking the stalemate. His decision to gamble on a one-on-one summit was his biggest risk as president. The announcement of the summit came at a time when Carter's popularity was at its lowest point. Both Sadat and Begin were under pressure not to deviate from their previously stated positions. The prospects for any compromise would be criticized all over the world.

During thirteen days of arduous negotiations, initiated in an atmosphere of gloom and mutual suspicion, Carter convinced the two leaders to accept two agreements that broke new ground. Essentially, the two accords represented agreements to agree, rather than an actual settlement of the difficult issues dividing the two nations. The two agreements reached dealt with the Sinai Peninsula and established a framework for settling the future of the West Bank and Gaza Strip.[16]

The negotiations proved to be difficult and time consuming. The December 17 deadline, specified in the agreement, soon passed and hopes for ratification faded. Sadat was pressured by the Arab nations not to accept a peace treaty with Israel at the expense of Palestinian rights. Begin's position in a coalition government was a great handicap.

The U.S. attempt to break this impasse through Secretary Cyrus Vance's shuttle diplomacy failed. A compromise proposal was submitted by the Carter administration. On March 15, 1979, the Egyptian cabinet voted unanimously to approve the peace treaty. The Israeli cabinet approved it on March 21. On March 26, the treaty was formally signed in Washington by Egypt and Israel, with Carter witnessing.

The treaty is a bilateral agreement between Egypt and Israel. Both were required to implement the framework established at Camp David in September 1978. A timetable was established for carrying out the provisions of the treaty. By the end of three years the final withdrawal of Israeli forces and civilian settlements from the Sinai would complete the timetable. This was done on April 25, 1982.

The above chronology of events leading to the final agreement is ample proof of the difficulties involved. There is no doubt that Carter's personal commitment and prestige was an important factor in the success of the peace negotiations. The threat of an Arab-Israeli war has been reduced because with Egypt at peace the remaining Arab states are not able to wage a war against Israel.

How far did the Arab use of oil as a political weapon precipitate this peace initiative in the Middle East? While the war was still in progress in 1973, the embargo spurred the United States to work for a peace settlement. As promised to Arab leaders, the United States moved in all seriousness to arrange a cease-fire and withdrawal of Israeli troops from the Suez Canal and part of the Golan Heights.

The signing of the Camp David agreement is not directly due to the impact of the oil embargo, although that could be argued at some length. The embargo was lifted five years before the signing of the Camp David agreement. Its impact was felt more in poor countries than in the United States. The United States, having suffered the least, has taken the initiative in the peace settlement, not because of fear of another embargo (adequate measures are being taken to prevent another embargo), but because of a desire to create a rapprochement between Israel and its Arab neighbors.

A major aim of the U.S. government was to encourage a peace settlement between Egypt and Israel, thus preventing Soviet inroads. But the treaty may ironically have the opposite effect. By insisting on the Camp David formula for peace, a formula which few Arabs could accept, the United States may have unintentionally encouraged a greatly expanded Soviet role in the Middle East.

The Camp David formula left open for future negotiation the settlement of the Palestinian questions, the future of the West Bank, the Golan Heights, and the Gaza Strip. The situation by the middle of 1982 raised no hope of real diplomatic effort in the future.

Achieving a peace settlement in the Middle East acceptable to all parties is a difficult task. Camp David was a beginning. Its momentum is lost. It can be revived if the Arabs and the Israelis want to avoid bloodshed and achieve territorial adjustment without war.

THE REAGAN PEACE PROPOSAL

Ronald Reagan put forward in September 1982 a peace plan for the Middle East which called for Palestinian self-government in the

West Bank and Gaza Strip in association with Jordan. Reagan ruled
out an independent Palestinian state in the territories occupied by
Israel in 1967. He also called for a freeze on new settlements there.
Reagan stated: "The Lebanon war, tragic as it was, has left us with a
new opportunity for Middle East peace."[17] Calling for a transition
period with full autonomy for Palestinians in the territories as specified in the Camp David accords, he added: "It is clear to me that
peace cannot be achieved by the formation of an independent state
in those territories. Nor is it achieved on the basis of Israeli sovereignty or on permanent control over the West Bank and Gaza." He
said although the United States would not support either alternative,
"it is the firm view of the United States that self-government by the
Palestinians of the West Bank and Gaza in association with Jordan
offers a just and lasting peace."[18] The Reagan peace plan also adds
that there should be a five-year transition period in which the Palestinians should demonstrate that they can manage their affairs and
that Palestinian autonomy poses no threat to Israel's security, because "America's commitment to the security of Israel is ironclad
and it should be universally accepted that the state of Israel deserves
unchallenged legitimacy within the community of nations." Reagan
also said that it was his position that in return for peace Israel should
withdraw from occupied areas on "all fronts, including the West
Bank and Gaza."[19]

Reagan sought to impress on high-level Arab delegates the view
that Jordanian King Hussein could play a key role in a renewed
Middle East peace plan. The United States believes that this could
be done if Arab states allow King Hussein to represent the Palestinians at a renewal of negotiations between Israel and Egypt. The
United States has remained adamant that it would hold talks with
the PLO only if it recognizes the Jewish state.

Israeli Deputy Premier David Levy, objecting to the proposal,
said that if implemented it would lead to the establishment of an
independent Palestinian state, even though this was not intended.
Levy said the new Reagan proposal would seriously threaten the
security and existence of Israel. Israel was particularly upset by
Reagan's suggestion of a link between Palestinian autonomy and
Jordanian monarchy whereby Israel would be required to make a
deal with Palestinians. Israeli Prime Minister Menachem Begin im-
pressed upon U.S. leaders Israel's belief that the Camp David autonomy plan was the best vehicle for settling the Palestinian issue.

Though Begin renewed the invitation for Jordan's King Hussein to join Palestinian autonomy talks, he said that any plan that placed the West Bank and Gaza outside the confines of Israel gives consent to the establishment of a Palestinian state.

Reagan's call for an end to Jewish settlements as part of his peace proposal was followed by Israel's announcement that it would set up 13 new settlements in the occupied Arab territories. Israeli troops in Lebanon also started building the first settlement in the occupied Lebanese Bekka Valley. Israeli Defense Minister Ariel Sharon announced on September 4, 1982 that if Lebanon did not sign a peace treaty with the Jewish state, Israel would retain portions of South Lebanon ranging 40–50 square kilometers. Sharon, after holding talks with the U.S. secretary of state, George Shultz, in Washington, stated: "Jordan is a Palestinian state." The Reagan administration disagreed with the Israeli minister's characterization of Jordan as a Palestinian state.

Israel is facing a serious crisis in its relations with Washington. It is seen as a crisis of confidence as well as policy. The Reagan administration acknowledges having consulted three Arab governments, Egypt, Saudi Arabia, and Jordan, before putting forward the plan, while Israel had been excluded from prior consultation.[20] The Reagan plan suggesting giving the Arabs the right to vote for the West Bank Council was also unacceptable to Israel.

The Reagan proposal is generally considered more sympathetic toward the Arabs than any earlier American proposals. Reagan formed what he called "a clear sense of America's position on the key issues." Stressing the United States' total commitment to Israel's security, he laid down a markedly changed policy toward the occupied areas. Reagan declared that in return for peace, the withdrawal position of Resolution 242 applies to all parts. He noted that the five-year transition period specified in the Camp David agreement following elections would prove that the Palestinians can run their own affairs and as such Palestinian autonomy poses no threat to Israel.

SOVIET GOALS

Soviet activities in the Middle East at first often appear to be ad hoc responses to (1) fortuitous circumstances such as the disruptions that have accompanied the process of Western decolonization,

(2) Western blundering in the area that all but drove Arab leaders such as Egypt's Nasser into the Soviet camp, and (3) extraregional military strategic considerations involving the Soviet Union's position vis-à-vis the NATO countries. The two broad Soviet concerns in the Middle East are defense of the Soviet homeland against an attack from the West, and expansion and consolidation of the Soviet role as a world power.

The Soviet strategy seems to be governed by the following constraints: (1) avoiding direct confrontation with the United States; (2) avoiding the formal commitment of Soviet forces to the region short of some critical situation such as in Afghanistan; and (3) avoiding two-front confrontations, that is, simultaneous military challenges from the local nations and the West. Within these constraints Soviet goals in the Middle East appear related to Soviet defense of the homeland against possible Western incursions.

The main tactic by which the Soviet Union has sought to achieve its strategic aims in the Middle East seems to be the support of guerrilla groups and regimes against the major Western powers. The principal instruments of this support are arms transfers, economic assistance, and declarations of political support for Arab causes. The other principal support has been the deployment of the Soviet navy to neutralize the Sixth Fleet's ability to intervene in local conflicts along the Mediterranean.

The Soviet penetration of the Arab Middle East has had far-reaching repercussions for inter-Arab conflicts and the Arab-Israeli conflict. First, after 1955 the Soviet Union attempted to raise its stake in the area by providing Arab states with an alternative source of arms. Second, until this time Moscow had maintained strict neutrality in the Arab-Israeli conflict. From the mid-1950s the Soviet Union began to cast an occasional vote in the United Nations for the Arab side instead of abstaining, as had been Moscow's custom in the early 1950s. The arms deal polarized the Arab states between Cairo and Baghdad and helped launch Nasser into world prominence as the leader of the dominant political force in the Arab world.

The recent catalyst for the Soviet Union was provided by the events in Afghanistan. The Soviet policy is based on the assumption that what is good for the United States is bad for the Soviet Union, and vice versa. It is Soviet policy to oppose U.S. policy. The diplomacy of two superpowers is reactive. When one of them acts, the other reacts. This action-reaction dichotomy operates much better in a region like the Middle East, which plays one against the other, each attempting to further its own interests. The Middle Easterners also contribute to the polarization and division of the area along cold-war dimensions.

The Soviets have repeatedly emphasized that the international environment has undergone substantial change in recent years, that the role of the West is diminishing and a new international force is emerging. If nothing else, the Soviets have shown in recent years that they have both the ability and the willingness to provide effective military support to their friendly regimes and movements. The success of the regimes in maintaining power in Afghanistan, Syria, Iraq, and South Yemen is attributable to Soviet military assistance and support. The image of the Soviet Union as equal or even militarily superior to the United States may well cause Middle Eastern governments to work out deals with the Soviet Union.

SOVIET INTEREST

The Soviet Union has a 3,000-mile border with the Middle East, which runs along the Black Sea, the Turkish and Iranian frontiers, and across Afghanistan to the Pamirs where the Soviet Union, China, and Afghanistan meet. The Soviet frontier with the Middle East is the only place apart from its Norwegian and Finnish borders where the Soviet Union adjoins the noncommunist world.[21] The Soviet Union is also sensitive to Middle Eastern affairs because its border areas with the Middle East have a predominantly Muslim population. The Soviet Union itself can become a major threat to the Middle East because it can easily deploy its land-based forces into the region, and has the capacity to maintain its momentum, as in Afghanistan.

The Soviet penetration of the Middle East began with the Egyptian arms deal in 1955. It started to a large extent because of the creation of the Central Treaty Organization (CENTO). Slowly, "the Arab world was certain to become a focus of Moscow's attention as Soviet interest shifted increasingly to the developing nations. The pace of Soviet activity was however accelerated by American manoeuvers in the Northern tier."[22]

The Soviet Union bypassed the northern tier countries of Turkey, Iran, and Pakistan and chose to concentrate on the Arab core of the Middle East. Its method was not to threaten but to woo the Arabs with the aid of satellites and a vigorous aid-and-trade policy.[23] It established such contacts with Egypt, Syria, South Yemen, and, since the revolution of 1958, with Iraq. The prestige gained during the Suez crisis further helped the Soviet Union strengthen its ties with the Arabs. By the end of the 1950s the Soviet Union had been successful in defusing the CENTO and denying the West a monopoly of influence in the Arab world. These ties were again strengthened after 1963. The Soviet Union became involved in the Palestinian issue, which was by then central in Arab politics. In the years following the 1967 war, the Soviet Union tried to rebuild Arab morale and made efforts to "prevent Israel's military victory from becoming a political victory as well."[24] The essential first step in Soviet calculations was a prompt rebuilding of Arab armed strength. By the end of 1967, a significant portion of Arab losses was replaced with up-to-date weaponry and an estimated $2.5 billion in arms.[25] The Soviet Union also got involved in Egypt's defense in a combat capacity.

The Soviet position on the Arab-Israeli dispute after 1967 was reflected in United Nations Resolution 242, which stated that Israel must be made to surrender Arab lands seized in the war, but it must be coerced through sanctions and diplomacy, not by force. The Soviets tried creating a parity of strength between the Arabs and the Israelis. They felt this would facilitate a peaceful solution to the crisis. However, not all Arabs agreed with Moscow's essentially moderate position. Algeria was critical of the Soviet position and Soviet-Syrian relations were disrupted for several years. The Palestinian guerrillas challenged Soviet strategists in a most serious manner. While Soviet press commentaries supported the guerrillas behind the scene, Soviet diplomats pressured Arab governments to curb the guerrillas and bring their activities under army discipline.

In the first half of the 1960s, the Soviet Union took steps to improve its ties with Turkey, Iran, and Pakistan. These countries and Afghanistan received, from 1963 to the end of 1969, approximately $1.2 billion in Soviet credit.[26] This amounted to more than a third of all Soviet aid pledged to developing nations during this period. Annual Soviet trade with the northern tier during these years averaged one-third of its total trade with the Third World (excluding India and Egypt). More than a third of the Soviet state visits during these years were to the northern tier. These overtures were motivated by a number of factors, including Moscow's desire to neutralize the tilt of these nations toward the West. Their moves were facilitated by an overall change in the power relationship from cold war to détente. These policies also came after the Sino-Soviet split and before the beginning of Soviet moves to circle China.

The Soviet buildup in the Mediterranean began at a time when the U.S. Sixth Fleet was already there. Soviet naval presence increased after the 1967 war. By 1970 there were 35 to 50 vessels regularly in the Mediterranean, at times as many as 75. This buildup was not related to any specific objective in the Middle East, but was part of a general development of naval strength throughout the world.

The Soviet naval buildup in the Indian Ocean began in 1967 with visits by ocean research units. By the next year a small Soviet squadron was in operation. By 1970, there were seven surface ships, four submarines, and one Polaris-type submarine.

The Soviet Union suffered a number of reverses in the Middle East. President Sadat abrogated the 1972 Treaty of Friendship and Cooperation with the Soviet Union. Syria and Iraq, while maintaining a friendly attitude toward Moscow, have slowly moved away from what was formerly a rather close association. Syrian disenchantment began when the Soviet Union opposed intervention in the Lebanese civil war. The 1975 defeat of Kurdish nationalists diminished Iraq's heavy reliance on the Soviet Union.[27]

The Soviet Union also improved its relationship with the PLO. Arafat paid his first visit to Moscow in 1970 as guest of the Soviet Afro-Asian Solidarity Organization. He received a state welcome from the Soviets in 1972. In 1977, a joint communiqué referred to the Geneva Conference as the only legitimate forum for conducting negotiations on the Arab-Israeli conflict and declared that PLO participation at Geneva was indispensable. Full diplomatic recognition of the PLO came during Arafat's visit to Moscow in October 1981.

By the summer of 1979, the Soviet Union became alert to the reality that a pro-Soviet ideologue, by the impetuous imposition of socialism, was turning Afghanistan from a traditionally friendly country to a hostile neighbor.[28] Therefore, the Soviet Union advised the Taraki government to slow reform. In September 1979, when Taraki returned from Moscow, he was overthrown by Hafizullah Amin. The Soviet Union then "saw no alternative but to plan a massive intervention which could simultaneously remove Amin, neutralize the Afghan army and tackle the intensifying insurgency."[29]

In Afghanistan, near the Iranian border, the Soviet Union has built an airbase at Shindand. From Shindand, the gulf is within striking range. After the invasion of Afghanistan nearly every airstrip in Afghanistan has been improved for Soviet use.

The Soviet Union may have overestimated its ability to bring the situation in Afghanistan to a quick and decisive conclusion, but the invasion was a low-risk military venture given the West's general lack of interest in the country. The invasion has catalyzed U.S. and regional concerns about Soviet political intentions. As a result, the Kremlin cannot assume that future political and military actions—particularly against such front-line states as Iran or Saudi Arabia—would be low-risk ventures. In the aftermath of the Soviet invasion, the United States would have preferred to have received more regional political and military support for its actions. Nevertheless, in terms of political, economic, and military importance, those Middle Eastern states favorably inclined toward the West (Egypt, Saudi Arabia, Oman, United Arab Emirates, Qatar, Bahrain, and Kuwait) far outnumber the states favorably inclined toward the Soviet Union (South Yemen, Afghanistan, and Syria). Thus, it would not be hard to imagine how a Soviet strategist may very well conclude that it is the United States, not the Soviet Union, that has more political maneuvering room in the area and, as a result, a better opportunity to swing the region's military balance in its favor.

The lack of real friends and allies in the region is an important Soviet constraint. As Alvin Rubinstein has argued, "with the possible exception of the PDRY (Peoples Democratic Republic of Yemen or South Yemen), none of the countries of the region wants to see a consolidated Soviet presence."[30] Some nations are willing to use Soviet arms and equipment to pursue their own nationalistic aspirations, but the political leverage that Moscow has gained has been modest at best.

The Soviet political system also offers few attractive features for Arab nations. The regimes of the Middle East are predominantly monarchical and Islamic. As a result, they have few long-term commonalities with communism or the Soviet Union. While Moscow's declaratory positions on antiimperialism, freedom, and democracy fall upon receptive ears, the reality of Moscow's actions in the region and within the Soviet Union gives many nations a negative impression of the Soviet Union.

With regard to the PDRY, its army is functioning under the joint command of Yemenis and Soviets. The Soviet naval base on Perim Island in the Bab-el-Mandab Strait has been expanded. An airfield at Bir Fadhl has also been expanded where Soviet Twelfth Air Force Squadron is stationed.

SOVIET OIL DILEMMA

Will energy needs push the Soviet Union toward the Persian Gulf oil? If the problem were a shortage, then there would be good reason to believe that future Soviet actions in the Middle East may be driven by economic interests. However, the Soviet dilemma is not a result of a lack of oil. It is the ability of the Soviet Union to meet compelling requirements as production begins to level off. Or, put another way, the Soviet dilemma has to do with how it can fulfill its domestic needs and also be a consistent supplier to Eastern Europe and other buyers of Soviet energy. As Thane Gustafson has recently argued, Soviet energy concerns are "more than a crisis of production; issues of consumption, distribution, and substitution of fuels are key elements also."[31] Therefore, Soviet oil concerns are part of a larger issue. There are tradeoffs that Moscow can and has already made to deal with its problems. How successful it will continue to be in the future is an open question, largely determined by domestic decisions made on questions related to resource allocation, investment priorities, energy supplies provided to Eastern and Western Europe, and development of alternative energy supplies in both the Soviet Union and Eastern Europe.[32]

This suggests that Soviet economic interests (including energy concerns) will play a role, but probably not a determining one, in Soviet behavior in the Third World. They will reinforce or undermine decisions made for other reasons.

IDEOLOGICAL CONSIDERATIONS

The Soviet ideological commitment to the Third World has undergone significant permutations, but its interest in the area is strongly rooted in history. Lenin in his 1916 work, *Imperialism, the Highest Stage of Capitalism*, saw the preconditions for a global socialist revolution inexorably tied to the developing nations. The imperialist powers' competition for colonies would inevitably lead to conflict and wars among themselves. This would advance the progress of socialism by hastening the demise of capitalism.

Leonid Brezhnev's approach toward the Third World has been more pragmatic, less simplified, more cautious, and clearly more rational.[33] Nevertheless, the ideological underpinnings and the conceptual importance of the Third World as part of the world revolutionary movement in the struggle against imperialism have remained largely intact during Brezhnev's rule. As Brezhnev told the Twenty-sixth Party Congress in February 1981: the Soviet Union "will continue to pursue consistently the development of cooperation between the USSR and the liberated countries in an effort to consolidate the alliance between world socialism and the national-liberation movement."[34]

Ideologically, Mikhail Gorbachev also sees a natural alliance between the Soviet Union and Third World nations. As one observer has said, "an unchanging Soviet ideological objective is to help forge that linkage."[35] To the degree that some Third World nations share a similar perception of linkage with the Soviet Union, Soviet ideological interests are achieved, and in meeting them other more important geopolitical and strategic interests are served.

SOVIET CAUTION

Significant gaps in our knowledge of the Soviet Union adversely impact on our ability to make definitive judgments about Soviet behavior. Generalizations about how the Soviet Union has pursued its interests and objectives suggest insight into Soviet strategy which will facilitate our subsequent discussion of Soviet constraints.

First, despite a vocal school of thought in the United States which argues that the Soviet Union believes it can fight and win a nuclear war, superpower conflict avoidance has guided Soviet actions.[36] It is one thing to want to be prepared to win a nuclear war if such an event is forced upon the Kremlin; it is quite another matter to initiate actions which might lead to a nuclear confrontation. Whether it was Korea in the 1950s, Cuba in 1962, Vietnam during the 1960s and 1970s, or the Middle East in 1973, Moscow has acted with extreme caution when a possibility existed that Soviet and U.S. forces might become engaged in combat.

Second, the Kremlin is quite sensitive to the fact that the appearance of military power has positive virtues and that military factors play an important role in reaching strategic nuclear parity. This has reduced the possibility of U.S. military intervention in parts of the Third World and has forced the United States to deal with the Soviets on an equal political and military basis. Soviet leaders believe that the growth of their total military power has permitted them to pursue a more active role in the world and expand Soviet influence. They see military strength as a critical element for expanding Soviet influence and consolidating present and past gains.

Third, despite its concerns about avoiding superpower conflict, the Kremlin has been willing to use and threaten with military force. As Afghanistan indicates, Moscow has even been willing to use a unilateral military invasion of a non-Soviet bloc country to secure its interests. However, in the final analysis, the Kremlin has not hastily resorted to the use of military force in the Third World or the Middle East. It has used military force when it believed that it could do so cheaply and when the risk of U.S. military counteraction was small.[37] In Afghanistan, Moscow may have miscalculated on how fast it could solve the problem and the actual extent of U.S. and world political reaction. Soviet advisers who counseled the Kremlin were right about the most crucial item: U.S. interests and objectives were not strong enough for it to risk superpower confrontation over Afghanistan.

Fourth, as Roger Kanet has argued, no "single unified policy toward the Third World" can accurately characterize Soviet behavior in the post-World War II period.[38] When asked if the Soviet Union is driven by offensive, hostile motives or if Moscow is defensive and reactive, the only accurate and reliable answer can be yes and no. The answer depends upon the circumstances, events, conditions, and who is answering the question.

OPPORTUNITIES AND ADVANTAGES

In its competition with the United States, a combination of conditions in the Middle East provides the Soviet Union with particular opportunities and advantages. The most often cited Soviet advantage is proximity. The Soviet Union is nearer to the region than is the United States. It also has contiguous borders with major parts of the region. The Soviet Union does not have to worry about creating a military presence in the region to show its interest or commitment. As Shahram Chubin has said, the Soviet Union "is in the region and unable to get out."[39] The geographic facts mean Moscow casts a long political shadow over the Middle East. As a result, the Soviet invasion of Afghanistan did not add a new dimension to the Middle Eastern nations' strategic considerations, particularly those states nearest the Soviet Union. They have always had to shape their policies and actions with an eye over their shoulder, and consider what their northern neighbor's reaction might be. The invasion of Afghanistan heightened and intensified already existing concerns. The primary opportunity that proximity provides the Soviet Union is that regional nations might become politically neutralized, frightened away from independent behavior, or be forced to seek accommodation with Moscow because of the Soviet Union's physical presence.

Soviet military forces in the region are possibly the second most mentioned advantage. For a Middle Eastern military contingency the Kremlin could draw upon 25 ground force divisions in the north Caucasus, Transcaucasus, and Turkistan military districts. There is one air tactical army in Turkistan and another in the Transcaucasus military district. Initially, they could provide between 450 and 600 aircraft to support a Soviet military operation in the Middle East. Finally, the Soviet Indian Ocean squadron, augmented with ships from the Pacific and possibly northern fleets, would present an immediate Soviet naval threat in the Middle East.

Regional political instability also provides the Soviets with advantages and opportunities. In an area of the world where the political status quo always seems to be in jeopardy, Moscow has the opportunity to achieve indirect benefits from events which it often neither initiates nor controls. For example, the indigenously sparked Iranian revolution, which led to the fall of the shah of Iran, made the United States politically and militarily vulnerable in the Middle East.

Although Moscow has not yet been very successful in exploiting these new opportunities, Iran's political disintegration, the collapse of the military, the lack of central leadership, the vehemently anti-U.S. attitude of the country, the revival of separatist ambitions among Iran's ethnic minorities, and the reappearance of the Tudeh party provide Moscow with opportunities that were not available under the shah.[40]

The unique conditions which led to the fall of the Pahlavi government cannot be re-created in other areas of the world. Where else in the Middle East does such a dynamic and respected religious leader as Ayatollah Khomeini exist who can lead a movement that cuts across economic, social, and class lines? Iran, however, should not be treated as an isolated example. The conditions which galvanized the Iranian radical left, middle class, extreme right, and fundamentalist clerics were internal problems of modernization: unfilled expectations, disparity of wealth, destruction of traditional values, corruption, a gap between economic reality and expectations, restricted participation in the process of government in a society which was monarchical and authoritarian, and a government that was weak given its fragile domestic political consensus.

The conditions—maybe not to the same degree—also exist in other important Middle Eastern nations. For example, Saudi Arabia must walk a fine line in its efforts to modernize, to protect traditional Islamic values and achieve a smooth integration of the large number of Western-educated students who in the future are supposed to hold important jobs in the bureaucracy and military. If the Saudis are successful, events like the 1979 attack on the Grand Mosque at Mecca will continue to be isolated incidents. However, if they are not successful in walking the tightrope, domestic instability will present Moscow with opportunities.

A legitimate troubling factor for U.S. policymakers is that the Saudi situation is not unique. Traditional ethnic problems in Iraq and Iran and large foreign work forces in Qatar, Kuwait, and the United Arab Emirates will be convenient targets of the Soviet Union when these countries face future domestic conflicts and crises. Suffice it to say, in regions like the Middle East, where the status quo seems to be in jeopardy, there are obvious advantages and opportunities favoring the Soviet Union.

Traditional regional rivalries endemic to the Middle East also provide Moscow with opportunities to exploit. Moscow's military power and its willingness to be a reliable and consistent supplier of military equipment is a primary instrument for gaining an opening into many nations. Without the existing regional rivalries, the Soviet Union would lack its most important means of access. While numerous studies indicate that arms sales frequently do not lead to direct influence over a recipient's actions, the Soviet Union would lose its most important entrée if regional rivalries did not exist.[41]

The unresolved Arab-Israeli issue also is a plus for the Soviet Union. Most Middle Eastern nations are not directly concerned with this issue. Strong U.S. support for Israel, which many Arabs believe is out of proportion to the United States' interests in the region, is an irritant to United States-Arab relations. The Arab-Israeli conflict provides radical Arab states with a rallying cause. It also threatens the political stability of Saudi Arabia by exposing the royal family to criticism and unwanted confrontation with radical elements within and outside Saudi Arabia. The Arab-Israeli issue also provides the Soviets with opportunities to influence some nations.

As John Campbell says: "the urge for nonalignment in the region is real, for it stems from experience of peoples long subject to domination by outside." While it is easy for the United States to argue that Soviet proposals for nuclear-free zones in the Middle East, renunciation of spheres of influence, recognition of countries' sovereignty over their natural resources, no outside support for separatist movements aimed at partitioning developing countries, and respect for nonalignment and developing countries' territorial integrity are not serious Soviet negotiating positions, "such proposals have a potential appeal to the peoples of the region."[42] A failure to recognize this appeal opens the door for Soviet propaganda victories, which in the large picture are not all that important, and to regionally imposed constraints upon U.S. potential and military actions because Washington is perceived as insensitive to Middle East needs, aspirations, and threats.

LEBANON

Israel's 1982 invasion of Lebanon began a series of events which changed the dimensions of the problem for all concerned. The active parties were Israel, the PLO, and Syria. The superpowers reacted to events, making decisions to encourage and support, to discourage and restrain, to stop the fighting or let it go on, each with the idea of furthering its own interests, maintaining allies and influence, and reducing the killing of innocent victims, mostly Lebanese civilians and Palestinian refugees. The role of the United States was more prominent than that of the Soviet Union. The Soviets, in the twilight of the Brezhnev era, were not looking for adventure in the Middle East but were clearly stung by events in Lebanon. They obviously did not wish to intervene militarily on behalf of the PLO or Syria or provoke hostilities between Syria and Israel. Their decision was to speed the arming of Syria, a well-established response to similar situations, and reaffirm Soviet backing for Syria.

Soviet support for the Arab cause has not been as strong as has been U.S. support for Israel's security. The Soviet Union will have to be on the defensive as long as it stays in Afghanistan. The United States will probably keep on playing with Soviet and Iranian threats to strengthen its foothold in the region, at least until CENTCOM becomes fully operational and facilities in the Middle East are developed to U.S. taste. In order to counteract CENTCOM and reassert its influence in the region, the Soviet Union would perhaps make some moves to woo the Arabs as it did in the mid-1950s after the formation of CENTO. There is a possibility that Moscow will then use more direct means as it did in Afghanistan. In that case, Iran will probably be the first target of superpower rivalry in the near future.

GAINS AND LOSSES

The United States, like the Soviet Union, finds that its access to military facilities in the Middle East and the Third World is dependent upon the vagaries of local developments. Although there has been an increase in U.S. arms to individual Arab states in recent years, the increase has been quite modest in comparison with the increase in Soviet arms. The United States desires to gain access to military facilities and support countries which see themselves threatened by Soviet-supported states.

To a very great extent both U.S. and Soviet policies in the Middle East have been characterized by concern for the activities of each other. Soviet policy has been motivated by a desire to gain advantages in the global competition with the United States, while U.S. policy has usually consisted of a reaction to Soviet policy. Both superpowers operate on the assumption that a gain for one represents an automatic loss for the other. There is little mutual interest. In the economic field, although Soviet trade with Middle East states has expanded, it remains low. Even states such as Syria, Iraq, and South Yemen, which have close ties with the Soviet Union, maintain many of their commercial ties with the West. On the other hand, U.S. trade is growing—especially imports of oil and exports of strategic materials.

Although similar to the Soviet objectives, U.S. objectives are broader. Containment of Soviet military and political influence in the Middle East, the acquisition of military facilities, and the securing of economic interests, such as oil, are of utmost importance.

In this superpower competition, the Middle Eastern states are forced to take sides. The introduction of sophisticated weapons by both superpowers into regional conflicts has led to an escalation of the level of conflict and in many cases increased expenditure for military security while neglecting economic development.

Neither the United States nor the Soviet Union has been able to make stable gains with the Middle East states, yet both continue to view the Middle East as an area of competition. Neither of the two superpowers has yet emerged as a winner, although it may be argued that the Middle East has been the loser. Superpower involvement in the Middle Eastern affairs has not been beneficial to the long-term interests of any nation.

The costs of broadening confrontation between the two major nuclear powers in the Middle East are great. For the two superpowers they include the large sums spent on arming and training the military of client states. For the Middle Easterners these costs involve an increasing focus on security problems at the expense of social and economic problems and issues. We are likely to see an escalation of regional conflicts, with destructive and disastrous consequences for the whole world.

NOTES

1. *Time*, October 25, 1982.

2. Abdul A. Said, ed., *America's World Role in the 70s* (Englewood Cliffs, N.J.: Prentice-Hall, 1970), p. 11.

3. Louis W. Koenig, ed., *The Truman Administration: Its Principles and Practices* (New York: New York University Press, 1956), p. 299.

4. Nadav Safran, "Engagement in the Middle East," *Foreign Affairs* 53, no. 1 (October 1974): 49–50.

5. Eugene V. Rostow, "A Basis for Peace," *New Republic* 172, no. 14 (April 1975): 12–13.

6. Safran, "Engagement," p. 54.

7. Ibid., p. 56.

8. George W. Ball, "The Looming War in the Middle East and How to Avert It," *Atlantic* 235, no. 1 (January 1975): 10–11.

9. Ibid., p. 6.

10. *Washington Post*, December 22, 1974.

11. Ibid., March 28, 1975.

12. Congressional Quarterly, *The Middle East: U.S. Policy, Israel, Oil and the Arabs* (1979): 35.

13. *New York Times*, March 17, 1977.

14. For a text of the joint statement, see Congressional Quarterly, *The Middle East: U.S. Policy, Israel, Oil and the Arabs*, loc. cit.

15. Ibid., p. 4.

16. Ibid.

17. *Durham Morning Herald*, September 2, 1982.

18. Ibid.

19. Ibid.

20. *Guardian Weekly* (London), September 3, 1982.

21. Fred Halliday, *Threat From the East* (Middlesex, England: Institute of Policy Studies, 1982), p. 43.

22. Charles B. McLane, *Soviet Middle East Relations* (London: Central Asian Research Center, 1973), p. 6.

23. George Lenczowski, *The Middle East in World Affairs* (Ithaca, N.Y.: Cornell University Press, 1962), p. 665.

24. McLane, *Soviet*, p. 9.

25. Ibid.

26. Ibid., p. 12.

27. O. M. Smolanski, "Soviet Policy in the Middle East," *Current History* 74, no. 433 (January 1978): 5.

28. Jagat S. Mehta, "Afghanistan: A Neutral Solution," *Foreign Policy* 47 (Summer 1982): 139–53.

29. Ibid.

30. Alvin Z. Rubinstein, "The Soviet Union and the Arabian Peninsula," *World Today* 35, no. 11 (November 1979): 452.

31. Thane Gustafson, "Energy and the Soviet Bloc," *International Security* 6, no. 3 (Winter 1981–82): 65.

32. Ibid., pp. 65–85.

33. Congressional Research Service, Library of Congress, *The Soviet Union and the Third World: A Watershed in Great Power Policy?* (A report to the Committee on International Relations, U.S. House of Representatives by the Senior Scientists Division) (Washington, D.C.: U.S. Government Printing Office, 1977), p. 2.

34. "The 22nd Congress of the Soviet Communist Party Proceedings and Related Materials," *Foreign Broadcasting Information Service: Daily Report, USSR and Eastern Europe* 7 (October 20, 1961): 85.

35. Congressional Research Service, Library of Congress, *Soviet Policy and the United States Response in the Third World* (A report prepared for the Committee on Foreign Affairs, U.S. House of Representatives) (Washington, D.C.: U.S. Government Printing Office, 1981), p. 30.

36. Richard Pipes, "Why the Soviet Union Thinks It Could Fight and Win a Nuclear War," *Commentary* 64, no. 1 (July 1977): 21–34.

37. Stephen S. Kaplan, ed., *Diplomacy of Power: Soviet Armed Forces as a Political Instrument* (Washington, D.C.: Brookings Institution, 1981), pp. 148–201.

38. Roger E. Kanet and Donna Bahry, eds., "The Soviet Union and the Developing Countries: Policy or Policies," in *Soviet Economic and Political Relations with the Developing World* (New York: Praeger, 1975), p. 10.

39. Shahram Chubin, "U.S. Security Interests in the Persian Gulf in the 1980s," *Daedalus* 109, no. 4 (Fall 1980): 48.

40. Alvin Rubinstein, "The Soviet Union and Iran Under Khomeini," *International Affairs* (London) 57, no. 4 (Autumn 1981): 599–617.

41. Alvin Rubinstein, *Red Star on the Nile: The Soviet-Egyptian Influence Relationship Since the June War* (Princeton, N.J.: Princeton University Press, 1977), pp. 5, 7, 13, 14, 18–19, 29–32.

42. John C. Campbell, "The Middle East: A House Containment Built on Shifting Sands," *Foreign Affairs* 60, no. 3 (1982): 624–25.

Postscript

Since the 1973–74 energy crisis, nations have been greatly influenced by their perceptions of Middle East oil power, perceptions based on the analyses and prognoses of the specialists in government, academia, research organizations, oil companies, banks, and others interested in oil politics and economics. These experts' projections of oil at $50 or more a barrel by 1986 are frightening.

Developments in the world oil market since the 1970s energy shocks have been major determinants of winners and losers. Contrary to the high hopes and expert opinions of oil-producing/exporting nations, the economic recovery in the West (particularly the United States) and oil demand remained stable. Further, to the surprise of both producers and consumers, demand waned in the United States and elsewhere while an oil glut developed in the world oil market. By winter of 1986 the demand for oil fell below the OPEC-imposed production limit of 16 million barrels daily (mbd).

When oil prices were multiplying during the 1970s many wondered how high they could go. In an effort to keep prices up, OPEC reduced production to less than 50 percent of its members' capacity. By March 4, 1986, the situation reversed: oil prices plummeted to $12 a barrel, the lowest since the embargo of 1973. The strategy backfired, the result of the price war begun by the Arab OPEC producers in the fall of 1985. Last December, when OPEC abandoned its strategy of defending prices to fight for a bigger share of the market, many people had hoped, somewhat unrealistically, that the price war implicit in the decision could somehow be avoided.

But Saudi Arabia, the largest OPEC oil producer, began to realize that it was being betrayed by Libya and other OPEC nations, which had exceeded their quotas. Angered, the Saudis increased their production from about 2 mbd to 4.5 mbd. These production and price changes will be felt economically and politically for a long time. The market instability may push energy costs even lower, precipitating further adjustments of policies between oil-producing and -consuming nations.

Where will the lower oil prices lead? Much depends on Saudi Arabia. Between 1981 and 1982 Saudi Arabia averaged production of 10 mbd. To produce at that level today would drive oil prices below $3 a barrel and cause a further drop in Saudi oil revenues, thereby deepening its current account deficit in the balance of payments.

It is interesting to note that Saudi export revenues dropped from a peak of $110 billion in 1981 to $33 billion in 1985. Saudi economy went into recession and its foreign assets dwindled by some $60 to $65 billion. Saudi Arabia's $55 billion in liquid assets virtually could be exhausted in two years, unless it increases oil earnings or cuts imports, which have been about $55 billion a year.

At what price then and at what level of output would the Saudis maximize their oil revenues and curb the erosion of their foreign assets? If the Saudis produced 8 mbd, the price would fall to $5 a barrel, and their current account deficit would rise to $32 billion. At the other extreme, if they produced only 1.5 mbd, the price would rise to $30 a barrel, but oil revenue would fall to $10.2 billion. Neither oil experts nor economists can solve this dilemma.

However, at a sustained price of $15 a barrel, the Saudis could produce 6 mbd, increase their oil revenues to $27.1 billion, and cut their current account deficit to $15.9 billion a year. At $22 a barrel, they could produce 4 mbd, but that would cut their oil revenues to $25.6 billion and lift their current account deficit to $17.4 billion.

It seems possible that the Saudis would do best to set a price close to $15 a barrel. But it is unlikely that they would let the official price of oil dip below $20 a barrel. This is simply because the lower prices would make it more difficult to set OPEC quotas that would generate acceptable revenues for the other producers in the existing world market, considering the abundance of non-OPEC sources.

The news of lower oil prices is met with mixed feelings. The good news is that oil-importing countries benefit immediately from lower prices. But how long will this last? Beyond 1990, the world oil outlook could change substantially.

In the absence of major new oil production outside the Middle East, and in the event that production in the United States and Britain declines during the 1990s, it follows that the Arab OPEC members would remain the main source of oil to the world. Any addition to oil supply from sources outside the Middle East is unlikely to offset declines elsewhere.

Some predictable changes are: (1) a loss of control of the oil market by the producers and further weakening of OPEC as an effective cartel; (2) conflict between rich and poor oil producers, over-populated and less-populated producers, and OPEC and non-OPEC producers; (3) direct deals between oil producers and traders that may bypass middlemen in the oil business; (4) lower inflation and lower interest rates in the world; (5) counterforces—tariffs and taxes—are likely to come into play in some of the major consuming nations; and (6) lower trade deficits in the United States.

By the mid-1990s the major oil producers will still have surpluses to export. But the major oil-consuming nations, particularly the United States, may not have enough oil to meet their needs. If OPEC can remain a unified and viable organization and again can dictate production and price figures, and if major oil-importing countries fail to develop energy alternatives to oil, then the bad news will surface. Then OPEC may take vengeance on the major oil consumers by creating an artificial supply crisis as it did in collaboration with its sister organization—the OAPEC—in 1973–74.

The current cheap oil could lull industrialized consumer nations into a false sense of security and into abandoning measures for energy independence. The resulting resurgence in the demand for oil could regain for OPEC a chance to control again the oil market in a few years. It could conceivably happen before the end of this century.

One last thought: The oil-consuming nations could be marking time until new sources of energy (nuclear, for instance) could bail them out.

Bibliography

This bibliography is selected mostly from books. It does not include the large number of journals and newspaper articles used.

Abdalati, Hammudah. *Islam in Focus.* Kuwait: International Islamic Federation of Student Organizations, 1978.

Abdulghani, Jasim M. *Iraq and Iran: The Years of Crisis*. London: Croom Helm, 1984.

Abir, Mordechai. *Oil, Power and Politics: Conflict in Arabia, the Red Sea and the Gulf*. London: Frank Cass, 1974.

Abrahamian, E. *Iran: Between the Two Revolutions*. Princeton, N.J.: Princeton University Press, 1982.

Amad, Adnan. *Israeli League for Human and Civil Rights: The Shahak Papers*. Beirut, Lebanon: Neebii, 1973.

Antonius, George. *The Arab Awakening*. Beirut, Lebanon: Khayats, 1955.

Arberry, A. J., ed. *Religion in the Middle East*. Cambridge: Cambridge University Press, 1969.

Ayoob, Mohammed, ed. *The Politics of Islamic Reassertion*. London: Croom Helm, 1981.

Bakhash, Shaul. *The Politics of Oil and Revolution in Iran*. Washington, D.C.: Brookings Institution, 1982.

Binder, Leonard. *The Ideological Revolution in the Middle East*. New York: John Wiley and Sons, 1964.

Bradley, C. Paul. *The Camp David Peace Process: A Study of Carter Administration Policies, 1979–1980*. Grantham, N.H.: Thompson and Rutter, 1981.

Brewer, Thomas L. *American Foreign Policy: A Contemporary Introduction*. Englewood Cliffs, N.J.: Prentice-Hall, 1980.

220

Burrell, R. M. *The Persian Gulf: The Washington Papers*, vol. 1. Beverly Hills, Calif.: Sage Publications, 1972.

Burrell, R. M., and Cottrell, Alvin J. *Iran, the Arabian Peninsula, and the Indian Ocean*. New York: National Strategy Center, 1972.

Chomsky, Noam. *Peace in the Middle East: Reflections on Justice and Nationhood*. New York: Vintage Books, 1974.

Choudhury, G. W. *Constitutional Development in Pakistan*. London: Longmans, Green, 1959.

Chubin, Sharam. *Security in the Persian Gulf: Domestic Political Factors*, vol. 1. Montclair, N.J.: Allanheld and Osmum, 1981.

Chubin, Sharam. *Security in the Persian Gulf: The Role of Outside Powers*, vol. 4. Montclair, N.J.: Allanheld and Osmum, 1982.

Chubin, Sharam. *Soviet Policy Towards Iran and the Gulf*. London: International Institute for Strategic Studies, 1980.

Clark, Wilson. *Energy for Survival: The Alternative to Extinction*. Garden City, N.Y.: Anchor Books, 1974.

Cohen, Aharon. *Israel and the Arab World*. Boston: Beacon Press, 1970.

Collins, John M. *U.S.-Soviet Military Balance, Concepts and Capabilities 1960–1980*. New York: McGraw-Hill, 1980.

Conant, Melvin A. *The Oil Factor in U.S. Foreign Policy 1980–1990*. Lexington, Mass.: Lexington Books, 1982.

Congressional Quarterly. *The Middle East: U.S. Policy, Israel, Oil and the Arabs*. Washington, D.C., 1979.

Cottrell, Alvin J., et al. *Sea Power and Strategy in the Indian Ocean*. Beverly Hills, Calif.: Sage Publications, 1981.

Cottrell, Alvin J., ed. *The Persian Gulf States*. Baltimore: Johns Hopkins University Press, 1980.

Cottrell, Alvin J., and Bray, Frank. *Military Forces in the Persian Gulf*. Washington Papers, vol. 6, no. 60. Beverly Hills, Calif.: Sage Publications, 1978.

Cudsi, S. Alexander, and Dessouki, A. E. H., eds. *Islam and Power*. Baltimore: Johns Hopkins University Press, 1981.

Curtis, Michael, ed. *Religion and Politics in the Middle East*. Boulder, Col.: Westview Press, 1981.

Darmstadter, Joel, Teitelbaum, Perry D., and Polach, Jaroslav G. *Energy in the World Economy*. Baltimore: Johns Hopkins University Press, 1971.

Davis, David Howard. *Energy Politics*. New York: St. Martin's Press, 1974.

Dawisha, Adeed. *Islam in Foreign Policy*. Cambridge: Cambridge University Press, 1983.

Deese, David A., and Nye, Joseph S., eds. *Energy and Security*. Cambridge, Mass.: Ballinger, 1981.

Dekmejian, R. Hrair. *Islam in Revolution: Fundamentalism in the Arab World*. Syracuse, N.Y.: Syracuse University Press, 1985.

Dessouki, A. H. *Islamic Resurgence in the Arab World*. New York: Praeger Publishers, 1982.

Donaldson, Robert H. *The Soviet Union in the Third World: Successes and Failures*. Boulder, Col.: Westview Press, 1981.

Donohue, John, and Esposito, J. L., eds. *Islam in Transition: Religion and Sociopolitical Change*. New York: Oxford University Press, 1982.

Dupree, Louis. *Afghanistan*. Princeton, N.J.: Princeton University Press, 1973.

Dupuy, Trevor N. *Elusive Victory: The Arab-Israeli Wars, 1947-1974*. New York: Harper and Row, 1978.

Enayat, Hamid. *Modern Islamic Political Thought*. London: Macmillan, 1982.

Erickson, Edward W., and Waverman, Leonard, eds. *The Energy Question: An International Failure of Policy*, vol. 1. Toronto: University of Toronto Press, 1974.

Esposito, John L. *Islam and Politics*. Syracuse, N.Y.: Syracuse University Press, 1984.

Esposito, John L., ed. *Islam and Development: Religion and Sociopolitical Change*. Syracuse, N.Y.: Syracuse University Press, 1980.

Evron, Yair. *The Middle East Nations, Superpowers and Wars*. New York: Praeger Publishers, 1973.

Ezzati, A. F. *The Concept of Leadership in Islam*. London: Muslim Institute, 1979.

Ezzati, A. F. *The Revolutionary Islam and the Islamic Revolution*. Tehran: Ministry of Islamic Guidance, Government of Iran, 1981.

Farid, Abdel Majid, ed. *Oil and Security in the Arabian Gulf*. London: Croom Helm, 1981.

Fisher, John C. *Energy Crisis in Perspective*. New York: John Wiley and Sons, 1974.

Fisher, Sydney N. *The Middle East: A History*. New York: Alfred A. Knopf, 1959.

Freedman, Robert O. *Soviet Policy Toward the Middle East Since 1970*. New York: Praeger Publishers, 1982.

Freedman, Robert O., ed. *World Politics and the Arab-Israeli Conflict*. New York: Pergamon Press, 1979.

Freeman, S. David. *Energy: The New Era*. New York: Vintage Books, 1974.

Gellner, Ernest. *Muslim Society*. Cambridge: Cambridge University Press, 1981.

Gibb, Hamilton A. R. *Modern Trends in Islam*. Chicago: University of Chicago Press, 1974.

Gibb, Hamilton A. R. *Mohammedanism*. New York: Oxford University Press, 1972.

Glassman, John D. *Arms for the Arabs: The Soviet Union and the War in the Middle East*. Baltimore: Johns Hopkins University Press, 1976.

Greene, Joseph N., Jr. *The Path to Peace: Arab-Israeli Peace and the United States, Report of a Study Mission to the Middle East*. Mount Kisco, N.Y.: Seven Springs Center, 1981.

Grummon, Stephen R. *The Iran-Iraq War: Islam Embattled*. New York: Praeger Publishers, 1982.

Guillaume, Alfred. *Islam*. New York: Penguin Books, 1954.

Gurion, David Ben. *Israel: A Personal History*. New York: Funk and Wagnalls, 1971.

Gurion, David Ben. *Rebirth and Destiny of Israel*. New York: Philosophical Library, 1954.

Halliday, Fred. *Threat From the East*. Middlesex, England: Institute of Policy Studies, 1982.

Hallwood, Paul, and Sinclair, Stuart. *Oil, Debt and Development: OPEC in the Third World*. London: Allen and Unwin, 1981.

Hammond, Allen L., Metz, William D., and Maugh, Thomas H. *Energy and the Future*. Washington, D.C.: American Association for the Advancement of Science, 1973.

Harkabi, Yehoshafat. *Arab Strategies and Israel's Response*. New York: Free Press, 1977.

Helms, Christine Moss. *Iraq: Eastern Flank of the Arab World*. Washington, D.C.: Brookings Institution, 1984.

Heper, Metin, and Israeli, Raphael. *Islam and Politics in the Middle East*. New York: St. Martin's Press, 1984.

Hertzberg, Arthur, ed. *The Zionist Idea: A Historical Analysis and Reader*. New York: n.p., 1959.

Herzl, Theodor. *The Jewish State*. New York: Scopus, 1943.

Hourani, Albert. *Arabic Thought in the Liberal Age, 1798–1939*. London: Oxford University Press, 1970.

Hourani, Albert. *The Emergence of the Modern Middle East*. Berkeley: University of California Press, 1981.

Hudson, Michael C. *Arab Politics: The Search for Legitimacy*. New Haven, Conn.: Yale University Press, 1977.

Hurewitz, J. C. *The Struggle for Palestine*. New York: W. W. Norton and Co., 1950.

Hurewitz, J. C., ed. *Oil, the Arab-Israeli Dispute and the Industrial World*. Boulder, Col.: Westview Press, 1976.

Hussain, Asaf. *Islamic Iran: Revolution and Counter-Revolution*. New York: St. Martin's Press, 1985.

Hussain, Asaf. *Political Perspectives on the Muslim World*. New York: St. Martin's Press, 1984.

Ibrahim, Saad Eddin. *The New Arab Social Order: A Study of the Social Impact of Oil Wealth*. Boulder, Col.: Westview Press, 1982.

Inglis, K. A. D., ed. *Energy From Surplus to Scarcity?* New York: John Wiley and Sons, 1974.

Insight Team of the Sunday Times. *Insight on the Middle East War*. London: Andre Deutsch, 1974.

International Institute for Strategic Studies. *Strategic Survey 1973*. London: IISS, 1974.

Ismael, Tareq Y. *The Arab Left*. Syracuse, N.Y.: Syracuse University Press, 1976.

Ismael, Tareq Y. *Iraq and Iran: Roots of Conflict*. Syracuse, N.Y.: Syracuse University Press, 1982.

Issawi, Charles, and Yeganeh, Mohammed. *The Economics of Middle Eastern Oil*. New York: Frederick A. Praeger, 1962.

Issawi, Charles, and Yeganeh, Mohammed. *Oil, the Middle East and the Third World*. New York: Library Press, 1972.

Jacoby, Neil H. *Multinational Oil: A Study in Industrial Dynamics*. New York: Macmillan Co., 1974.

Jansen, G. H. *Militant Islam*. London: Pan Books, 1979.

Jeffries, J. M. N. *Palestine: The Reality*. London: Longmans, Green and Co., 1939.

Jensen, W. G. *Energy and the Economy of Nations*. Oxfordshire, England: G. T. Fouls and Co., 1970.

Jiryis, Sabri. *The Arabs in Israel*. Beirut, Lebanon: Institute for Palestine Studies, 1969.

Jureidini, Paul A., and McLaurin, R. D. *Beyond Camp David: Emerging Alignments and Leaders in the Middle East*. Syracuse, N.Y.: Syracuse University Press, 1981.

Kedourie, Elie. *Islam in the Modern World*. New York: Holt, Rinehart and Winston, 1981.

Kerr, Malcolm H. *The Arab Cold War*. New York: Oxford University Press, 1971.

Key, Kherim K. *The Arabian Gulf State Today*. Washington, D.C.: Asia Research Center, 1974.

Khadduri, Majid. *War and Peace in the Law of Islam*. Baltimore: Johns Hopkins University Press, 1955.

Khalidi, Rashid, and Mansour, Camille, eds. *Palestine and the Gulf*. Beirut, Lebanon: Institute for Palestine Studies, 1982.

Khomeini, Ayatollah R. *Islamic Government*. New York: Manor Books, 1979.

Khouri, Fred J. *The Arab-Israeli Dilemma*. Syracuse, N.Y.: Syracuse University Press, 1968.

Koening, Louis W., ed. *The Truman Administration: Its Principles and Practices*. New York: New York University Press, 1956.

Koury, Enver M. *The Arabian Peninsula, Red Sea and Gulf: Strategic Consideration*. Hyattsville, Md.: Institute of Middle East and North African Affairs, 1979.

Kramer, Martin. *Political Islam: The Washington Papers*, vol. 8. Beverly Hills, Calif.: Sage Publications, 1980.

Lancaster, Lane W. *Masters of Political Thought*. London: George G. Harrap and Co., 1959.

Lapids, Ira M. *Contemporary Islamic Movement in Historical Perspective*. Berkeley: University of California Press, 1983.

Leeman, Wayne A. *The Price of Middle East Oil: An Essay in Political Economy*. Ithaca, N.Y.: Cornell University Press, 1962.

Lenczowski, George. *The Middle East in World Affairs*. Ithaca, N.Y.: Cornell University Press, 1968.

Lenczowski, George. *Oil and State in the Middle East*. Ithaca, N.Y.: Cornell University Press, 1960.

Lenczowski, George, ed. *The Political Awakening in the Middle East*. Englewood Cliffs, N.J.: Prentice-Hall, 1970.

Lenczowski, George. *Soviet Advances in the Middle East*. Washington, D.C.: American Enterprise Institute for Public Policy Research, 1971.

Lenczowski, George, ed. *United States Interests in the Middle East*. Washington, D.C.: American Enterprise Institute for Public Policy Research, 1968.

Lewis, Bernard, *The Arabs in History*. New York: Harper and Row, 1966.

Lewis, Bernard. *The Middle East and the West*. New York: Harper and Row, 1964.

Macrakis, Michael S., ed. *Energy: Demand, Conservation, and Institutional Problems*. Cambridge, Mass.: MIT Press, 1974.

Makdisi, Samir A. *Middle East Problem Paper*, vol. 12: *Oil Price Increases and the World Economy*. Washington, D.C.: Middle East Institute, 1974.

Mancke, Richard. *The Failure of U.S. Energy Policy*. New York: Columbia University Press, 1974.

Mansfield, Peter. *The Arab World: A Comprehensive History*. New York: Crowell, 1977.

McLane, Charles B. *Soviet Middle East Relations*. London: Central Asian Research Center, 1973.

McLaurin, R. D., Mughisuddin, Mohammed, and Wagner, Abraham R. *Foreign*

Policy Making in the Middle East: Domestic Influences on Policy in Egypt, Iraq, Israel, and Syria. New York: Praeger Publishers, 1977.

Mendenhall, George. *The Tenth Generation.* Baltimore: Johns Hopkins University Press, 1973.

Miller, Roger LeRoy. *The Economics of Energy: What Went Wrong and How We Can Fix It.* New York: William Morrow and Co., 1974.

Mitchell, Edward J. *U.S. Energy Policy: A Primer.* Washington, D.C.: American Enterprise Institute for Public Policy Research, 1974.

Monroe, Elizabeth. *The Changing Balance of Power in the Persian Gulf.* New York: American Universities Field Staff, 1972.

Moore, J. N., ed. *The Arab-Israeli Conflict.* Princeton, N.J.: Princeton University Press, 1975.

Mosley, Leonard. *Power Play: Oil in the Middle East.* Baltimore: Penguin Books, 1974.

Mughisuddin, Mohammed, ed. *Conflict and Cooperation in the Persian Gulf.* New York: Praeger Publishers, 1977.

Naff, Thomas, ed. *The Middle East Challenge, 1980–1985.* Carbondale, Ill.: Southern Illinois University Press, 1981.

Nakhleh, Emile A. *The Persian Gulf and American Policy.* New York: Praeger Publishers, 1982.

Nakhleh, Emile A. *The United States and Saudi Arabia.* Washington, D.C.: American Enterprise Institute for Public Policy Research, 1975.

Nasser, Gamal Abdel. *Speeches and Press Interviews, January–December 1963.* Cairo, Egypt: U.A.R. Information Department, 1964.

Nathan, Robert R., Gass, Oscar, and Creamer, Daniel. *Palestine: Problem and Promise.* Washington, D.C.: Public Affairs Press, 1946.

Niblock, Tim, ed. *Social and Economic Development in the Arab Gulf.* London: Croom Helm, 1980.

Norton, A. R. *Moscow and the Palestinians: A New Tool of Soviet Policy in the*

Middle East. Washington, D.C.: Center for Advanced International Studies, University of Miami, 1974.

Novik, Nimrod, and Starr, Joyce, eds. *Challenges to the Middle East Regional Dynamics and Western Security.* New York: Praeger Publishers, 1981.

Odum, Howard T. *Environment, Power, and Society.* New York: John Wiley and Sons, 1971.

O'Neill, Bard. *Armed Struggle in Palestine.* Boulder, Col.: Westview Press, 1979.

Pachachi, Nadim. *The Role of OPEC in the Emergence of New Patterns in Government-Company Relationships.* London: Royal Institute of International Affairs, 1972.

Pennar, Joan. *The U.S.S.R. and the Arabs: The Ideological Dimension.* New York: Crane, Russak, 1973.

Penrose, Edith T. *The Growth of Firms, Middle East Oil and Other Essays.* London: Frank Cass, 1971.

Peretz, Don. *Israel and the Palestine Arabs.* Washington, D.C.: Middle East Institute, 1958.

Peretz, Don. *The Middle East Today.* New York: Holt, Rinehart and Winston, 1965.

Piscatori, James P., ed. *Islam in the Political Process.* Cambridge: Cambridge University Press, 1983.

Plascov, Avi. *Security in the Persian Gulf: Modernization, Political Development and Stability.* Vol. 3, Institute for Strategic Studies. Montclair, N.J.: Allanheld and Osmum, 1982.

Polk, William R. *The Elusive Peace: The Middle East in the Twentieth Century.* New York: St. Martin's Press, 1980.

Pragner, Robert J. *American Policy in the Middle East 1960-1971.* Washington, D.C.: American Enterprise Institute for Public Policy Research, 1971.

Pragner, Robert J., and Tahtinen, Dale R. *Nuclear Threat in the Middle East.* Washington, D.C.: American Enterprise Institute for Public Policy Research, 1975.

Price, David Lynn. *Oil and the Middle East Security: The Washington Papers*, vol. 7. Beverly Hills, Calif.: Sage Publications, 1976.

Quandt, William B. *Decade of Decisions: American Policy Toward the Arab-Israeli Conflict, 1967–1976*. Berkeley: University of California Press, 1977.

Quandt, William B., Jabber, Fuad, and Lesch, Ann Mosely. *The Politics of Palestinian Nationalism*. Berkeley: University of California Press, 1973.

Rahman, Fazlur. *Islam*. Chicago: University of Chicago Press, 1979.

Ramazani, Rouhollah K. *The Northern Tier: Afghanistan, Iran, and Turkey*. Princeton, N.J.: D. Van Nostrand Co., 1966.

Ramazani, Rouhollah K. *The Persian Gulf: Iran's Role*. Charlottesville: University of Virginia Press, 1973.

Ramazani, Rouhollah K. *The United States and Iran: The Patterns of Influence*. New York: Praeger Publishers, 1982.

Randall, Jonathan C. *Going All the Way: Christian Warlords, Israeli Adventures and the War in Lebanon*. New York: Viking Press, 1983.

Record, Jeffrey. *The Rapid Deployment Force and U.S. Military Intervention in the Persian Gulf*. Cambridge, Mass.: Institute for Foreign Policy Analysis, 1981.

Reich, Bernard. *Quest for Peace: U.S. Israeli Relations and the Arab-Israeli Conflict*. Brunswick, N.J.: Transaction Books, 1977.

Ridgeway, James. *The Last Play: The Struggle to Monopolize the World's Energy Resources*. New York: E. P. Dutton and Co., 1973.

Rifai, Taki. *The Pricing of Crude Oil: Economic and Strategic Guidelines for an International Energy Policy*. New York: Praeger Publishers, 1974.

Rocks, Lawrence, and Runyon, Richard P. *The Energy Crisis*. New York: Crown Publishers, 1972.

Roi, Yaacov, ed. *The U.S.S.R. and the Muslim World*. London: Allen and Unwin, 1984.

Roosevelt, Kermit. *Arabs, Oil and History: The Story of the Middle East*. New York: Harper and Row, 1949.

Rouhani, Fuad. *A History of OPEC*. New York: Praeger Publishers, 1971.

Roumaini, Maurice M., ed. *Forces of Change in the Middle East*. Worcester, Mass.: Worcester State College Press, 1971.

Rubinstein, Alvin. *Red Star on the Nile: The Soviet-Egyptian Influence Relationship Since the June War*. Princeton, N.J.: Princeton University Press, 1977.

Rustow, Dankwart A. *A Middle Eastern Political System*. Englewood Cliffs, N.J.: Prentice-Hall, 1971.

Safran, Nadav. *From War to War: The Arab-Israeli Confrontation 1948-1967*. New York: Pegasus, 1969.

Safran, Nadav. *The United States and Israel*. Cambridge, Mass.: Harvard University Press, 1963.

Said, Abdul A., ed. *America's World Role in the 70s*. Englewood Cliffs, N.J.: Prentice-Hall, 1970.

Said, Edward W. *Covering Islam*. New York: Pantheon, 1981.

Salibi, Kamal S. *Crossroads to Civil War: Lebanon 1958-1976*. Delmar, N.Y.: Caravan, 1976.

Sayyegh, Kamal S. *Oil and Arab Regional Development*. New York: Frederick A. Praeger, 1968.

Schmidt, Dana Adams. *Armageddon in the Middle East*. New York: John Day Co., 1974.

Schurr, Sam H., Homans, Paul T., and Darmstadter, Joel. *Middle Eastern Oil and the Western World: Prospects and Problems*. New York: American Elsevier Publishing Co., 1971.

Shariati, A. *An Approach to the Understanding of Islam*. Iran: Shariati Foundation, 1979.

Spagnolo, J. P. *France and Ottoman Lebanon 1861-1914*. London: Ithaca Press, 1977.

Stocking, George W. *Middle East Oil: A Study in Political and Economic Controversy*. Kingsport, Tenn.: Vanderbilt University Press, 1970.

Stoddard, Philip H., Cuthell, David C., and Sullivan, Margaret W., eds. *Change and the Muslim World*. Syracuse, N.Y.: Syracuse University Press, 1981.

Stork, Joe. *Middle East Oil and the Energy Crisis*. New York: Monthly Review Press, 1975.

Taher-Kheli, Shirin, and Ayubi, Shaheen. *The Iran-Iraq War: New Weapons, Old Conflicts*. New York: Praeger Publishers, 1983.

Tahtinen, Dale R. *The Arab-Israeli Military Balance Today*. Washington, D.C.: American Enterprise Institute for Public Policy Research, 1973.

Tahtinen, Dale R. *Arms in the Persian Gulf*. Washington, D.C.: American Enterprise Institute for Public Policy Research, 1974.

Tanzer, Michael. *Energy Crisis: World Struggle for Power and Wealth*. New York: Monthly Review Press, 1974.

Taylor, Alan R. *Prelude to Israel*. Beirut, Lebanon: Institute of Palestine Studies, 1970.

Toynbee, Arnold J. *Civilization on Trial*. New York: Oxford University Press, 1948.

U.S. Department of State. *U.S. Policy Toward the Persian Gulf: Current Policy No. 390*. Washington, D.C.: May 10, 1982.

Vicker, Ray. *The Kingdom of Oil*. New York: Charles Scribner's Sons, 1974.

Vocke, Harald. *The Lebanese War: Its Origins and Political Dimensions*. London: C. Hurst and Co., 1978.

Voll, John Obert. *Islam: Continuity and Change in the Modern World*. Boulder, Col.: Westview Press, 1982.

Watt, W. Montgomery. *Islam and the Integration of Society*. London: Routledge and Kegan Paul, 1961.

Watt, W. Montgomery. *What is Islam?* New York: Longman, 1979.

Weizmann, Chaim. *Trial and Error*. New York: Harper and Row, 1949.

Whetten, Lawrence L. *The Soviet Presence in the Eastern Mediterranean*. New York: National Strategic Information Center, 1971.

Wise, George S., and Issawi, Charles, eds. *Middle East Perspectives: The Next Twenty Years*. Princeton, N.J.: Princeton University Press, 1981.

Yager, Joseph A., and Steinberg, Eleanor B. *Energy and U.S. Foreign Policy*. Cambridge, Mass.: Ballinger Publishing Co., 1974.

Yannacone, Victor, Jr., ed. *Energy Crisis: Danger and Opportunity*. St. Paul, Minn.: West Publishing Co., 1974.

Yorke, Valerie. *The Gulf in the 1980s: Chatham House Papers*. London: Royal Institute of International Affairs, 1980.

Young, Arthur N. *Saudi Arabia: The Making of a Financial Giant*. New York: New York University Press, 1983.

Zabih, Sepehr. *Iran Since the Revolution*. Baltimore: Johns Hopkins University Press, 1982.

Ziring, Lawrence. *The Middle East Political Dictionary*. Santa Barbara, Calif.: ABC-Clio Information Services, 1984.

Zureik, Elia T. *The Palestinians in Israel: A Study in Internal Colonialism*. London: Routledge and Kegan Paul, 1979.

Index

About the Author

Sheikh R. Ali is a Professor of Political Science at North Carolina Central University.

A prodigious writer, Dr. Ali has published in the area of oil politics, Islam, the Middle East, and the Third World. The author of seven books and over three dozen articles and reviews, he has published in many national and international journals, including the *American Political Science Review* and the *Middle East Review*. His forthcoming book is *Oil and Power: Political Dynamics in the Middle East* (Frances Publishers, London, 1987).

Dr. Ali holds a Ph.D. in International Studies from American University and has triple master's degrees in government, political science, and international relations. He was a Fulbright scholar at New York University from 1965 to 1967. A former diplomat based in Washington, Professor Ali is an internationally known authority on world oil politics and a top petrologist from the Third World.

A rapporteur for President Carter's Commission on Foreign Languages and International Studies in 1979, Dr. Ali is a member of the Republican Task Force and he has been awarded a Presidential Gold Medal of Merit by Ronald Reagan.